Principles and Applications of NanoMEMS Physics

MICROSYSTEMS

Series Editor
Stephen D. Senturia
Massachusetts Institute of Technology

OTHER BOOKS IN THE SERIES:

- **Optical Microscanners and Microspectrometers Using Thermal Bimorph Actuators**
  Series: Microsystems, Vol. 14
  Lammel, Gerhard, **Schweizer**, Sandra, **Renaud**, Philippe
  2002, 280 p., Hardcover, ISBN: 0-7923-7655-2
- **Optimal Synthesis Methods for MEMS**
  Series: Microsystems, Vol. 13
  Ananthasuresh, S.G.K. (Ed.)
  2003, 336 p., Hardcover, ISBN: 1-4020-7620-7
- **Micromachined Mirrors**
  Series: Microsystems, Vol. 12
  Conant, Robert
  2003, XVII, 160 p., Hardcover, ISBN: 1-4020-7312-7
- **Heat Convection in Micro Ducts**
  Series: Microsystems, Vol. 11
  Zohar, Yitshak
  2002, 224 p., Hardcover, ISBN: 1-4020-7256-2
- **Microfluidics and BioMEMS Applications**
  Series: Microsystems, Vol. 10
  Tay, Francis E.H. (Ed.)
  2002, 300 p., Hardcover, ISBN: 1-4020-7237-6
- **Materials & Process Integration for MEMS**
  Series: Microsystems, Vol. 9
  Tay, Francis E.H. (Ed.)
  2002, 300 p., Hardcover, ISBN: 1-4020-7175-2
- **Scanning Probe Lithography**
  Series: Microsystems, Vol. 7
  Soh, Hyongsok T., **Guarini**, Kathryn Wilder, **Quate**, Calvin F.
  2001, 224 p., Hardcover
  ISBN: 0-7923-7361-8
- **Microscale Heat Conduction in Integrated Circuits and Their Constituent Films**
  Series: Microsystems, Vol. 6
  Sungtaek Ju, Y., **Goodson**, Kenneth E.
  1999, 128 p., Hardcover, ISBN: 0-7923-8591-8
- **Microfabrication in Tissue Engineering and Bioartificial Organs**
  Series: Microsystems, Vol. 5
  Bhatia, Sangeeta
  1999, 168 p., Hardcover, ISBN: 0-7923-8566-7
- **Micromachined Ultrasound-Based Proximity Sensors**
  Series: Microsystems, Vol. 4
  Hornung, Mark R., **Brand**, Oliver
  1999, 136 p., Hardcover, ISBN: 0-7923-8508-X

Principles and Applications of NanoMEMS Physics

by

Héctor J. De Los Santos
NanoMEMS Research LLC,
Irvine, CA, USA

A C.I.P. Catalogue record for this book is available from the Library of Congress.

ISBN 10 1-4020-3238-2 (HB)
ISBN 13 978-1-4020-3238-7 (HB)
ISBN 10 0-387-25834-5 (e-book)
ISBN 13 978-0-387-25834-8 (e-book)

Published by Springer,
P.O. Box 17, 3300 AA Dordrecht, The Netherlands.

www.springeronline.com

Printed on acid-free paper

Printed in the Netherlands.

Este libro lo dedico a mis queridos padres y
a mis queridos Violeta, Mara, Hector F. y Joseph.

"Y sabemos que a los que aman a Dios todas las cosas les ayudan a bien,
esto es, a los que conforme a su propósito son llamados."

Romanos 8:28

CONTENTS

PREFACE

This book presents a unified exposition of the physical principles at the heart of NanoMEMS-based devices and applications. *NanoMEMS* exploits the convergence between nanotechnology and microelectromechanical systems (MEMS) brought about by advances in the ability to fabricate nanometer-scale electronic *and* mechanical device structures. In this context, NanoMEMS-based applications will be predicated upon a multitude of physical phenomena, e.g., electrical, optical, mechanical, magnetic, fluidic, quantum effects and mixed domain.

Principles and Applications of NanoMEMS Physics contains five chapters. Chapter 1 provides a comprehensive presentation of the fundamentals and limitations of nanotechnology and MEMS fabrication techniques. Chapters 2 and 3 address the physics germane to this dimensional regime, namely, quantum wave-particle phenomena, including, the manifestation of charge discreteness, quantized electrostatic actuation, and the Casimir effect, and quantum wave phenomena, including, quantized electrical conductance, quantum interference, Lüttinger liquids, quantum entanglement, superconductivity and cavity quantum electrodynamics. Chapter 4 addresses potential building blocks for NanoMEMS applications, including, nanoelectromechanical quantum circuits and systems (NEMX) such as charge detectors, the which-path electron interferometer, and the Casimir oscillator, as well as a number of quantum computing implementation paradigms, including, the ion-trap qubit, the NMR-qubit, superconducting qubits, and a semiconductor qubit. Finally, Chapter 5 presents NanoMEMS applications in photonics, particularly focusing on the generation, propagation, and detection of surface plasmons, and emerging devices based on them.

The book assumes a preparation at the advanced undergraduate/beginning graduate student level in Physics, Electrical Engineering, Materials Science, and Mechanical Engineering. It was particularly conceived with the aim of providing newcomers with a much needed coherent scientific base for undertaking study and research in the NanoMEMS field. Thus, the book takes great pains in rendering transparent advanced physical concepts and techniques, such as quantum information, second quantization, Lüttinger liquids, bosonization, and superconductivity. It is also hoped that the book will be useful to faculty developing/teaching courses emphasizing physics and applications of nanotechnology, and to Nanotechnology researchers engaged in analyzing, modeling, and designing NanoMEMS-based devices, circuits and systems.

ACKNOWLEDGMENTS

The idea for this book began to take shape upon meeting Mr. Mark de Jongh, Senior Publishing Editor of Springer, at the European Microwave Conference in Munich, Germany, in October, 2003. Unbeknownst to the author, Dr. Harrie A.C. Tilmans, of IMEC, Belgium, had recommended him to Mr. de Jongh as a potential author. Upon a "chance" encounter Mr. de Jongh introduced himself and suggested the writing of a book for (then) Kluwer. The author submitted the book proposal in late November, 2003 and received news of its acceptance soon thereafter, as Springer's Microsystems book series editor, Dr. Stephen D. Senturia,. had provided a "very positive and complementary report." Therefore, the author is pleased to acknowledge Dr. Tilmans, for bringing his name to Mr. de Jongh's attention, Dr. Senturia, for his positive recommendation of the book proposal, and Mr. de Jongh, for providing him with the opportunity to write the book. Furthermore, the author gratefully acknowledges Mr. Mark de Jongh and Ms. Cindy M. Zitter, his Senior Assistant, for their patience and understanding during the course of the work. The book cites more than 200 references. Access to these would not have been possible without the excellent assistance of Mr. Tim Lee, whom he gratefully acknowledges. Finally, the author gratefully acknowledges the understanding of his wife, Violeta, along the course of the project, as well as her excellent assistance in preparing the final camera-ready manuscript.

Héctor J. De Los Santos

Chapter 1

NANOELECTROMECHANICAL SYSTEMS

1.1 NanoMEMS Origins

The field of Nanotechnology, which aims at exploiting advances in the fabrication and controlled manipulation of nanoscale objects, is attracting worldwide attention. This attention is predicated upon the fact that obtaining early supremacy in this field of miniaturization may well be the key to dominating the world economy of the 21st century and beyond. NanoMEMS exploits the convergence between nanotechnology and microelectromechanical systems (MEMS) brought about by advances in the ability to fabricate nanometer-scale electronic and mechanical device structures. Indeed, the impact of our ability to make and control objects possessing dimensions down to atomic scales, perhaps first considered by the late Richard Feynman in his 1959 talk "There is Plenty of Room at the Bottom" is expected to be astounding [1], [2]. In particular, miniaturization, he insinuated, has the potential to fuel radical paradigm shifts encompassing virtually all areas of science and technology, thus giving rise to an unlimited amount of technical applications. Since high technology fuels the prosperity of the world's most developed nations, it is easy to see why the stakes are so high.

Progress in the field of miniaturization benefited from the advent of the semiconductor industry in the 1960s, and its race to increase profits through the downscaling of circuit dimensions which, consequently, increased the density and the yield of circuits fabricated on a given wafer area. This density, which derived from progress in photolithographic tools to produce the ever smaller two-dimensional patterns (device layouts) of an integrated circuit (IC), has increased since by more than seven orders of magnitude and has come to be captured by Moore's law: The number of components per chip doubles every 18 months [2]. The culmination of such miniaturization program, it is widely believed, is the demise of Moore's law, whose manifestation is already becoming apparent due to an increasing predominance of the quantum mechanical nature of electrons in determining the behaviour of devices with critical dimensions (roughly) below 100 nm.

This line of development is closely related to the field of quantum devices/nanoelectronics, which was prompted by the conception of a number of atomic-level deposition and manipulation techniques, in particular, molecular beam epitaxy (MBE), originally exploited to construct laboratory devices in which the physics of electrons might be probed and explored, following the discovery of electron tunnelling in heavily-doped *pn*-junctions [3]. Nanoelectronics did produce interesting physics, for instance, the discovery of Coulomb blockade phenomena in single-electron transistors, which manifested the particle nature of electrons, and resonant tunnelling and conductance quantization in resonant tunnelling diodes and quantum point contacts, respectively, which manifested the wave nature of electrons [4-6]. These quantum devices, in conjunction with many others based on exploiting quantum phenomena, generated a lot excitement during the late 1980s and early 1990s, as they promised to be the genesis for a new digital electronics exhibiting the properties of ultra-high speed and ultra-low power consumption [7-8]. While efforts to realize these devices helped develop the skills for fabricating nanoscale devices, and efforts to analyze and model these devices helped to develop and mature the field of mesoscopic quantum transport, the sober reality that cryogenic temperatures would be necessary to enable their operation drastically restricted their commercial importance. A few practical devices, however, did exert commercial impact, although none as much as that exerted by silicon IC technology, in particular, heterojunction bipolar transistors (HBTs), and high-electron mobility transistors (HEMTs), which exploit the conduction band discontinuities germane to heterostructures, and modulation doping to create 2-D electron confinement and quantization, respectively, and render devices superior to their silicon counterparts for GHz-frequency microwave and low-transistor-count digital circuit applications [9-14].

The commercial success of the semiconductor industry, and its downscaling program, motivated emulation efforts in other disciplines, in particular, those of optics, fluidics and mechanics, where it was soon realized that, since ICs were fundamentally two-dimensional entities, techniques had to be developed to shape the third dimension, necessary to create mechanical devices exhibiting motion and produced in a batch planar process [15]. These techniques, which included surface micromachining, bulk micromachining, and wafer bonding, became the source of what are now mature devices, such as accelerometers, used in automobile air bags, and pressure sensors, on the one hand, and a number of emerging devices, such as, gyroscopes, flow sensors, micromotors, switches, and resonators, on the other. Coinciding, as they do, with the dimensional features germane to ICs, i.e., microns, these mechanical devices whose behavior was controlled by electrical means, exemplified what has come to be known as the field of microelectromechanical systems (MEMS).

Three events might be construed as conspiring to unite nanoelectronics and MEMS, namely, the invention of a number of scanning probe microscopies, in particular, scanning tunneling microscopy (STM) and atomic force microscopy (AFM), the discovery of carbon nanotubes (CNTs), and the application of MEMS technology to enable superior RF/Microwave systems (RF MEMS) [16-18]. STM and AFM, by enabling our ability to manipulate and measure individual atoms, became crucial agents in the imaging of CNTs and other 3-D nanoscale objects so we could both "see" what is built and utilize manipulation as a construction technique. CNTs, conceptually, two-dimensional graphite sheets rolled-up into cylinders, are quintessential nanoelectromechanical (NEMS) devices, as their close to 1-nm diameter makes them intrinsically quantum mechanical 1-D electronic systems while, at the same time, exhibiting superb mechanical properties. MEMS, on the other hand, due to their internal mechanical structure, display motional behavior that may invade the domain of the Casimir effect, a quantum electrodynamical phenomenon elicited by a local change in the distribution of the modes in the zero-point fluctuations of the vacuum field permeating space [19-21]. This effect which, in its most fundamental manifestation, appears as an attractive force between neutral metallic surfaces, may both pose a limit on the packing density of NEMS devices, as well as on the performance of RF MEMS devices [22].

In the balance of this chapter, we present the fundamentals of the fabrication techniques which form the core of NanoMEMS devices, circuits and systems.

1.2 NanoMEMS Fabrication Technologies

NanoMEMS fabrication technologies extend the capabilities of conventional integrated circuit (IC) processes, which are predicated upon the operations of forming precise patterns of metallization and doping (the controlled introduction of atomic impurities) onto and within the surface and bulk regions of a semiconductor wafer, respectively, with the performance of the resulting devices depending on the fidelity with which these operations are effected. Excellent books on IC fabrication, giving in-depth coverage of the topic, already exist [23] and the reader interested in process development is advised to consult these. The exposition undertaken here is cursory in nature and only aims at providing an understanding of the fundamentals and issues of present and future NanoMEMS fabrication technologies.

1.2.1 Conventional IC Fabrication Processes

Conventional IC processes are based on photolithography and chemical etching, and are synthesized by the iterative application to a wafer of a cyclic sequence of steps, namely: Spin-casting and patterning, material deposition, and etching. The salient elements of these steps are presented in what follows.

1.2.1.1 Spin-Casting

The first step (after thoroughly cleaning the wafer), in defining a pattern on a wafer, is to coat it with a photoresist (PR), Figure 1-1, a viscous light-

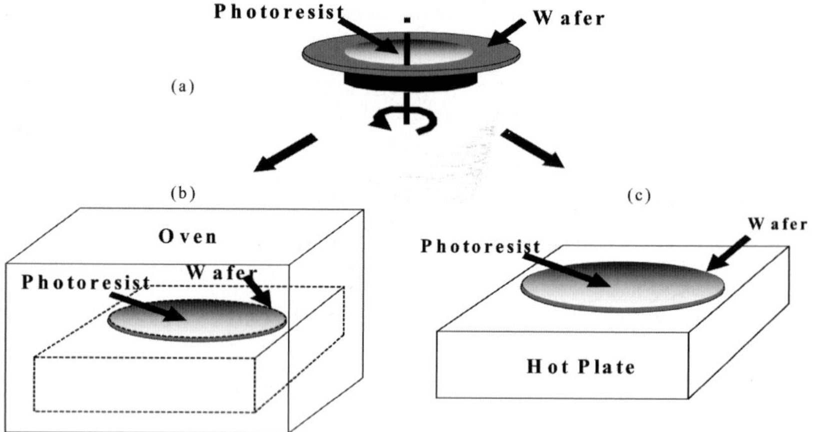

Figure 1-1. Coating wafer with photoresist. (a) Spin-casting. (b) Soft-bake in oven. (c) Hard bake in hot plate.

sensitive polymer whose chemical composition changes upon exposure to ultraviolet (UV) light. The process of applying the PR to the wafer in order to achieve a uniform thickness is called spin-casting, and usually involves the following steps: 1) Pouring a few drops of the PR at the wafer center; 2) Spinning the wafer for about 30 seconds once it reaches a prescribed rotational speed of several thousand revolutions-per-minute; and 3) Baking it at temperatures of several hundred degrees Celsius to produce a well-adhered solvent-free dry layer. The resulting PR film thickness is inversely proportional to the square root of the rotational speed, and directly proportional to the percent of solids in it. Determining these parameters is one of the first steps in developing a process.

1.2.1.2 Wafer Patterning

Once a uniform solid PR layer coats the wafer, this is ready for patterning. This is accomplished by interposing a glass mask, which contains both areas that are transparent and areas that are opaque, between a UV source and the PR-coated wafer. As a result, selective chemical changes are effected on the PR in accordance with the desired pattern, Figure 1-2. When it

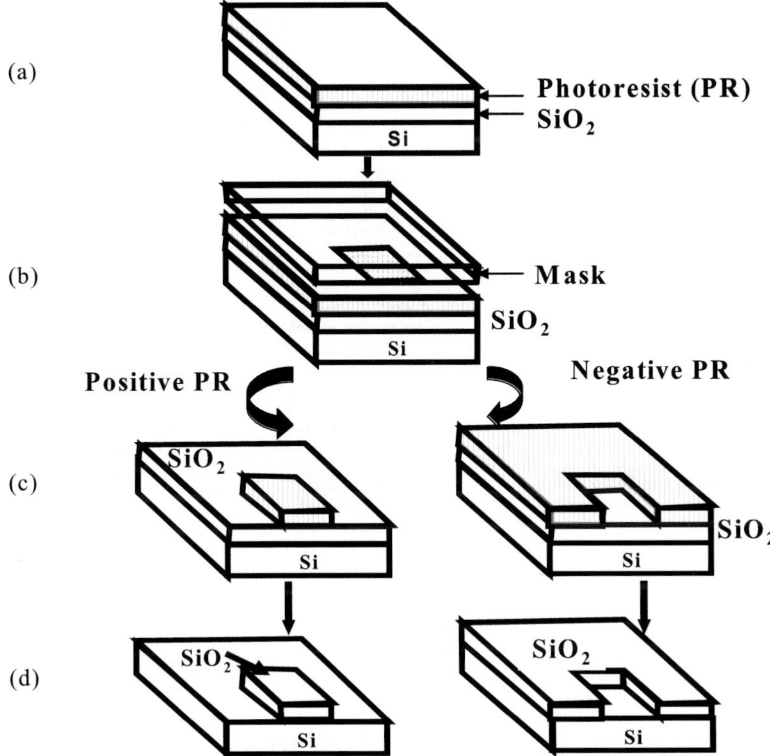

Figure 1-2. Wafer patterning with positive and negative photoresists. (*After* [24]).

is desired that the created pattern be identical to that in the glass mask, a positive PR, which hardens when exposed to UV light, is employed. Otherwise, when it is desired that the created pattern be the negative of that in the mask, a negative PR is employed. In the former case, UV exposure hardens the PR, whereas in the latter, UV exposure weakens the PR. Thus, subsequently, when the UV-exposed wafer is etched, the weakened parts of the PR will be dissolved and the desired pattern revealed. There are two techniques to dissolve the PR, namely, wet and dry etching. These are presented next.

1.2.1.2.1 Lithography

The highest resolution (minimum size) and quality of the pattern to be defined on a wafer depends on how well the mask image is transferred to the PR. Image formation, in turn, is determined by the lithographic process and type of PR employed. The lithographic process can make use of an optical source, an electron beam source, or an X-ray source for creating the desired pattern on the wafer. In this section we deal with the first and the last approaches.

Optical lithography, Figure 1-3, may be employed in conjunction with

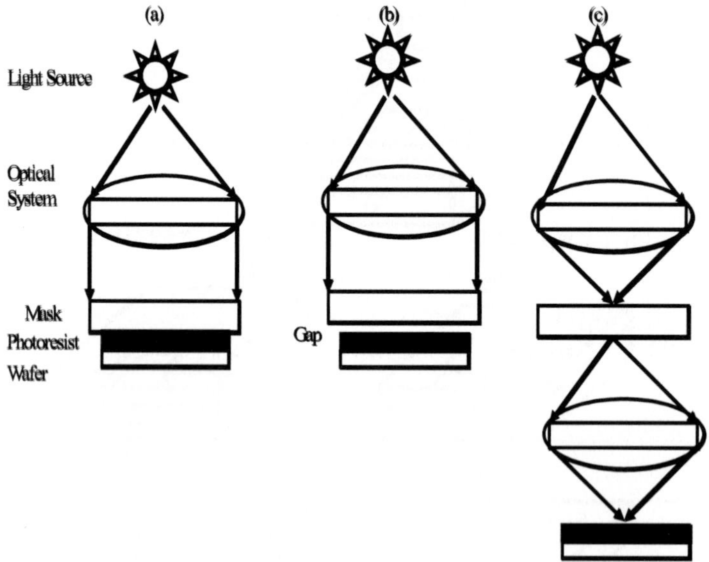

Figure 1-3. Sketches of common approaches to optical lithography. (a) Contact printing. (b) Proximity. (c) Projection. (*After* [23]).

either, contact printing, in which the image is projected through a mask that is in intimate contact with the wafer, or proximity printing, in which the image is projected through a mask separated by $\sim 10-25\mu m$ from the wafer, or projection printing, in which the mask is separated many centimeters away from the underlying wafer. Because, the contact and proximity approaches are prone to suffer from dust particles present between the mask and the PR, the projection approach is preferred for creating nanoscale-feature patterns. The resolution of a good projection optical lithography system is given by $0.5(\lambda/NA)$, where λ is the exposure wavelength and NA is the numerical aperture of the projection optics, at a depth of focus capability of $\pm\lambda/2(NA)^2$ [23]. The highest resolution of optical photolithography appears to be about 250nm-100nm for production devices, down to 70nm for laboratory devices, and is set by diffraction, i.e., at smaller sizes features become blurred. Overcoming these technical issues, which involves developing smaller wavelength light sources and optics, is difficult. Thus, the cost of optical lithography production equipment capable of reaching resolutions below 100 nm, is deemed by industry as prohibitive [24].

X-ray lithography, see Fig. 1-4, utilizing the low energy of soft x-rays at wavelengths between 4 and 50 Å, is relatively impervious to scattering effects.

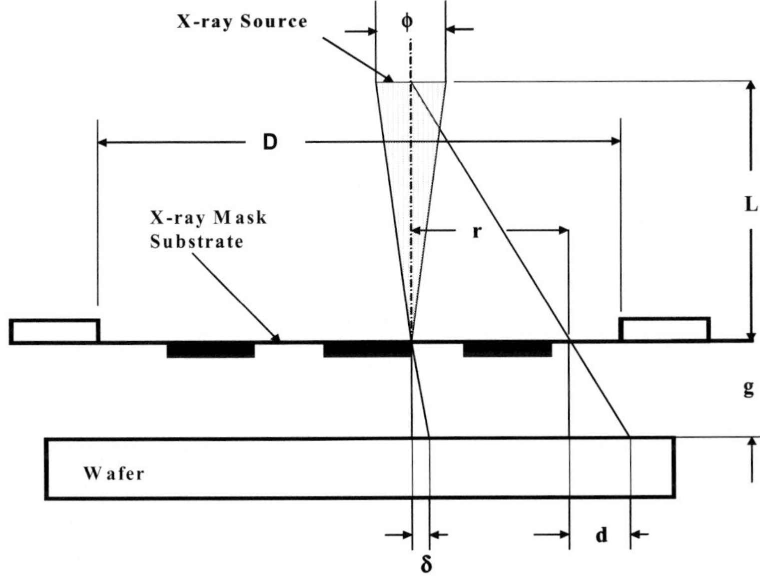

Figure 1-4. Sketch of factors eliciting geometrical limitations in x-ray lithography. Typical values for the geometrical parameters are: $\phi = 3mm$, $g = 40\mu m$, $L = 50cm$, $r = 63mm$. (*After* [23].)

This makes them amenable for use in exposing thick PRs which, because of their low absorption, can penetrate deeply and produce straight-walled PR images with high fidelity. Because of difficulty in creating optical elements at these wavelengths, however, the method of image projection employed is proximity printing through a mask containing x-ray absorbing patterns. The mask is separated from the PR-wafer a distance of just about $25\mu m$, but since dust particles with low atomic number do not absorb x-rays, no damage is caused to the pattern. Despite the potential for highest resolution germane to x-ray lithography, two factors have been identified as potentially limiting it. Both factors originate in geometrical aspects of the illumination. In particular, there is the possibility that a significant penumbral blur $\delta = \phi\, g/L$ be introduced on the position of the resist image by the extended point source of diameter ϕ located a distance L above a mask separated from the wafer a distance g. Also, a potential for lateral magnification error is present, due to the divergence of the x-ray from the point source and the finite mask to wafer separation. Accordingly, images of the projected mask are shifted laterally by an amount $d = r\, g/L$.

Even with perfect resolution, pattern formation quality depends on how the PR responds to the impinging lightwave or electron beam. This is addressed next.

1.2.1.2.2 Photoresist

The mechanism for image transfer to the PR involves altering its chemical or physical structure so the exposed area may subsequently be easily dissolved or not dissolved. According to the previous sections, pattern formation is effected on optical resists, electron beam resists, or x-ray resists.

Optical lithography resists may be negative or positive. The fundamental difference, in terms of how they affect the resolution of the image transferred, is rooted in their chemical composition.

In the negative resist, which combines a cyclized polyisopropene polymer material with a photosensitive compound, the latter becomes activated by the absorption of energy with wavelengths in the 2000- to 4500-Å range. The photosensor acts as an agent that causes cross linking of the polymer molecules by transferring to them the received energy. As a result of the cross linking, the molecules' molecular weight increases and this elicits their insolubility in the developing system. The highest resolution limit of a negative PR derives from the fact that during development the exposed (cross linked) areas swell, whereas the unexposed low molecular weight areas are dissolved. The minimum resolvable feature when using a negative resists is typically three times the film thickness [23].

In response to light the positive resist, which also contains a polymer and a photosensitizer, the latter becomes insoluble in the developer and, thus, prevents the dissolution of the polymer. Since the photosensitizer precludes the developer from permeating the PR film, no film swelling is produced and a greater resolution is possible [23].

Electron beam lithography also utilizes negative and positive resists. In a negative resist, the electron beam prompts cross-linking of the polymer, which results on increased molecular weight, increased resistance to the developer, and swelling during development. A common negative resist used with electron beam lithography is COP, poly (glycidylmethacrylate-co-ethyl acrylate), which renders a resolution of $1\mu m$. In a positive resist, the electron beam causes chemical bond breaking, reduced molecular weight, and reduced resistance to dissolution during development. Common positive resists used with electron beam lithography include poly(methyl methacrylate) (PMMA) and poly(butane-1 ketone) (PBS), which render a resolution of $0.1\mu m$.

X-ray lithography also utilizes negative and positive resists, in particular, COP, PBS and PMMA with resolution similar to that stated above is obtained.

1.2.1.3 Etching

Defining the desired pattern on the PR coating the wafer is crucial. The pattern fidelity is defined its selectivity and aspect ratio, Figure 1-5.

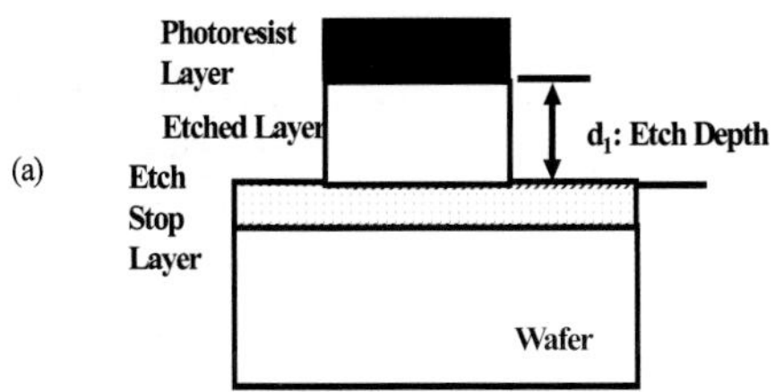

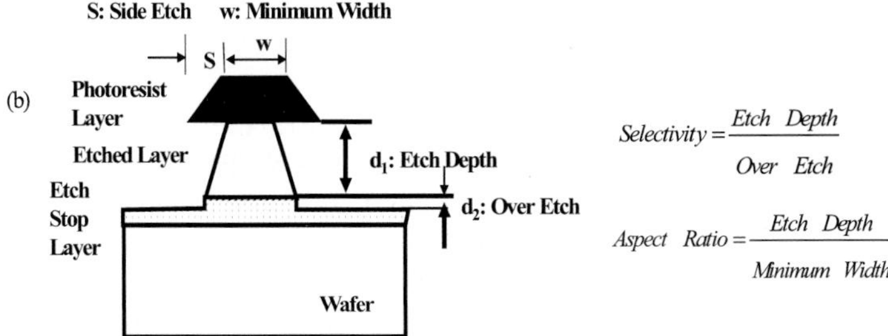

Figure 1-5. Pattern transfer definition. (a) Ideal. (b) Realistic. (*After* [25].)

It is seen in this figure that the fidelity of the pattern transferred is function of how precisely the resulting width of the etched layer resembles that of the PR pattern, as quantified by the selectivity and aspect ratio. Accordingly, four scenarios may be envisioned, Figure 1-6, which reflect the relative strength with which the etchant attacks the PR, the etched material, and the etch stop. In particular, it may be surmised from Figure 1-6(d) that the minimum width of a pattern, i.e., how narrow it may be, is limited by the lithography process to define the pertinent width in the PR and the resulting degree of undercut of the PR mask. Thus, etchants producing isotropic profiles (ones in which the vertical and horizontal etching rates are equal), are not amenable to pattern the narrowest features. In general, the results depend on a number of factors controlling the etching chemical reaction, such as temperature and mixing conditions, whether or not the etching agent employed is in the liquid or gaseous state, how well the PR adhered to the wafer during spin-casting. In the next section we address two of the most important factors, namely, the state of the etchant.

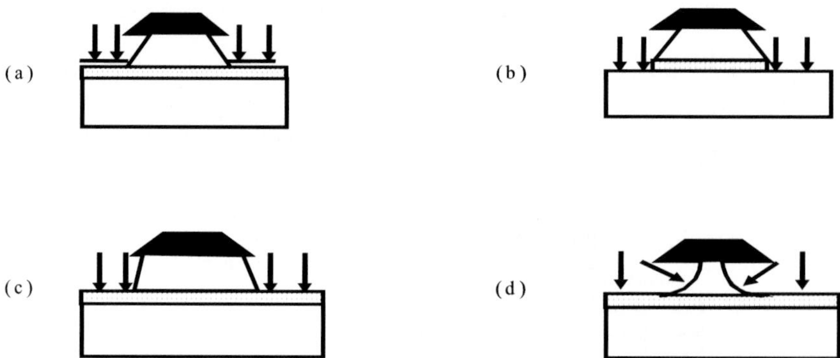

Figure 1-6. Etching characterization. (a) Over Etch<<Etch Depth→Selective. (b) Over Etch~Etch Depth→Non-selective. (c) Side Etch<<Etch Depth. (d) Side Etch~Etch Depth. (*After* [25].)

1.2.1.3.1 Wet Etching

In this approach to dissolve the weakened PR, the patterned wafers are immersed in a liquid chemical etchant, Figure 1-7. The etched profile may be isotropic or anisotropic depending of the wafer orientation. If this is amorphous, an isotropic profile will result, i.e., the horizontal and vertical etching rates are similar. Otherwise, if it is single-crystal, an anisotropic profile may result. A number of chemicals employed to effect anisotropic etching in silicon are in use. These include tetramethylammonium hydroxide (TMHA), potassium hydroxide (KOH), and ethylene diamine pyrochatecol

(EDP). Detailed experiments to elucidate the mechanism responsible for anisotropic etching have been undertaken [23]. The fundamental principle behind anisotropic etching appears to be this: when different crystal planes possess different atomic densities, those planes with greater density will etch at a slower rate than those with lower atomic density.

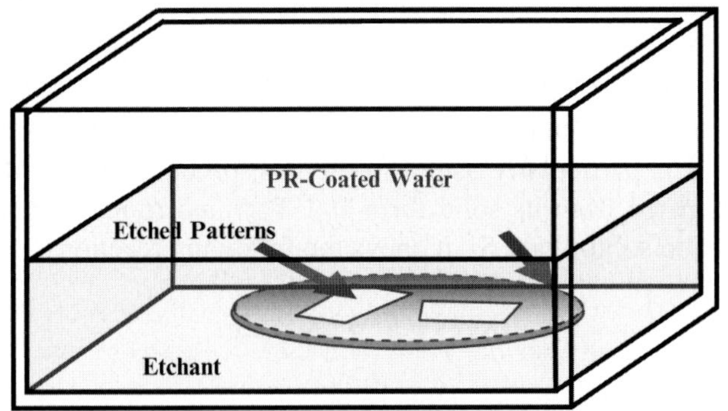

Figure 1-7. Etching of wafer immersed in liquid chemical solution.

An exhaustive compilation of chemical reactions for pertinent etching chemicals/wafer materials has been published by Williams and Muller [29]. Table 1-1 below gives some of typical etched material/etching solvent pairs.

Table 1-1. Wet etching targets and solvents

Etched Material	Etching Solvent
Silicon	KOH, TMAH, EDP
Silicon oxide	HF
Silicon nitride	H_3PO_4
Aluminum	H_3PO_4

When it comes to creation of free-standing structures via surface micromachining techniques (described below), wet etching is accompanied by various drawbacks. For instance, the surface tension exerted on the delicate free-standing structures by the fluid's hydrodynamic forces may preclude their complete release, or may even break them. Dry etching techniques, circumvent these drawbacks and are discussed next.

1.2.1.3.2 Dry Etching

In this approach, shown in Figure 1-8, a gas/vapor or plasma is used as a source of reactive atoms that dissolve the weakened PR. Typical matching pairs of etched material and etching gas used in IC fabrication are shown in Table 1-2.

Table 1-2. Etched material-etching gas pairs.

Silicon or Polysilicon	SF_6, CF_4
Silicon dioxide	CHF_4/H_2
Silicon nitride	CF4/O2
Aluminum	BCl_2

Two fluorine-containing gases have been recently adopted for dry etching processes, namely, *Xenon difluoride*, XeF_2 [30] and *Boron Fluoride*, BrF_3 [30]. XeF_2 enables an isotropic dry-etch process for silicon, which is very selective to aluminum, silicon dioxide, silicon nitride and photoresist. The XeF_2 gas is particularly useful in the post-processing of CMOS ICs. It can be sublimated from its solid form at 1 Torr and room temperature and, when applied to solid-phase Si, it obeys the following reaction:

$$2XeF_2 + Si \rightarrow 2Xe + SiF_4$$

XeF_2 etching of Si achieves high selectivity with a number of masking materials, such as, SiO_2, Si_3N_4, Al, PR, and phosphosilicate glass (PSG), at etching rates ranging from $1-3\mu m/min$ to as high as $40\mu m/min$ [30], and is characterized by the production of measurable amounts of heat. When in the presence of water or vapor, XeF_2 reacts with them to form HF. In terms of its potential application to nanostructure formation, XeF_2 etching has the drawback that the resulting surfaces tend to have a granular finish with a feature size of about $10\mu m$.

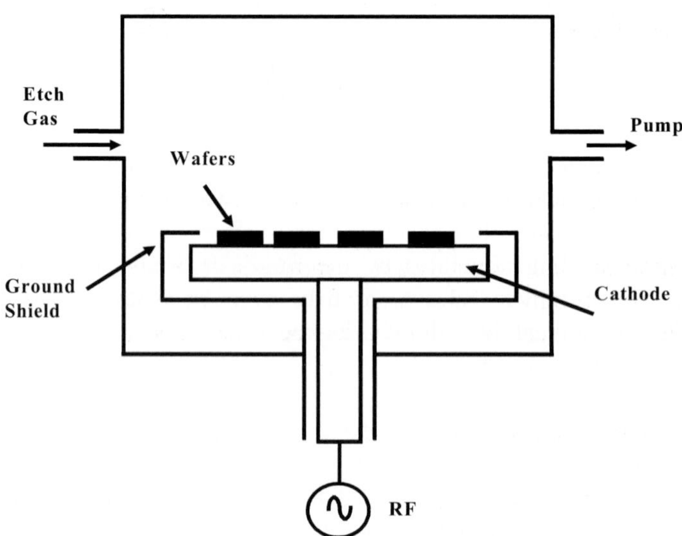

Figure 1-8. Etching of wafer immersed in plasma

BrF_3 on the other hand, enables isotropic etching of Si with masking materials such as Al, Au, Cu, Ni, PR, SiO_2, and Si_3N_4, while achieving surface finish feature size of 40-150nm. Dry etching, it may be concluded, is not amenable to creating nanostructures.

1.2.1.4 Chemical Vapor Deposition

The result of patterning a wafer is to render some areas of its surface bare to receive the deposition of various atomic species, while preventing such deposition in other areas. Chemical vapor deposition (CVD) is one of the techniques utilized to introduce atoms into the exposed wafer areas and, for silicon wafers, entails the dissociation of gasses, such as silane, SiH_4, arsine (AsH_3), phosphine (PH_3), and diborane (B_2H_6), on the wafer surface at high temperatures, usually in the 450-800°C range. The chamber containing the wafers during the deposition, Fig. 1-9, is usually held at pressures between 0.1 and 1Torr, and the resulting properties of the deposited materials varies.

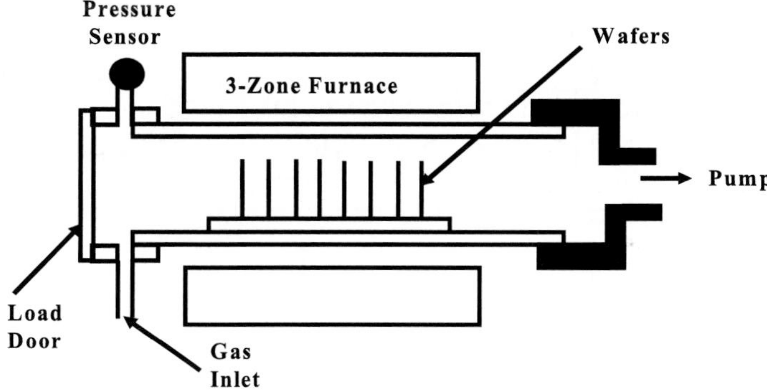

Figure 1-9. Schematic of hot-wall, reduced pressure CVD reactor.

For instance, under appropriate parameters of temperature, deposition rate, and crystallinity of the wafer, the deposited material may *grow epitaxially*, i.e., maintaining the same crystallographic nature of the substrate wafer, or become polycrystalline, i.e., exhibiting an agglomeration of randomly oriented crystallites. In the context of silicon processes, typical materials deposited via CVD include: polycrystalline silicon (polysilicon), silicon dioxide (SiO_2), and stoichiometric silicon nitride (Si_xN_y), to thicknesses $\sim 2\mu m$. The most common reactions for depositing these materials are shown in Table 1-3.

Table 1-3. Common CVD reactions and deposition temperatures for pertinent materials. [24]

Product	Reactants	Deposition temperature (°C)
Silicon dioxide	$SiH_4+CO_2+H_2$	850-950
	$SiCl_2H_2+N_2O$	850-900
	SiH_4+N_2O	750-850
	SiH_4+NO	650-750
	$Si(OC_2H5)_4$	650-750
	SiH_4+O_2	400-450
Silicon nitride	SiH_4+NH_3	700-900
	$SiCl_2H_2+NH_3$	650-750
Polysilicon	SiH_4	600-650

An alternate method to effect material deposition on a wafer while avoiding the high temperatures required in a CVD reactor is to utilize a hot-wall plasma deposition reactor, Fig. 1-10. In this approach, the wafers are oriented vertically in contact with long alternating slabs of graphite or aluminum electrodes inside a quartz tube heated by a furnace.

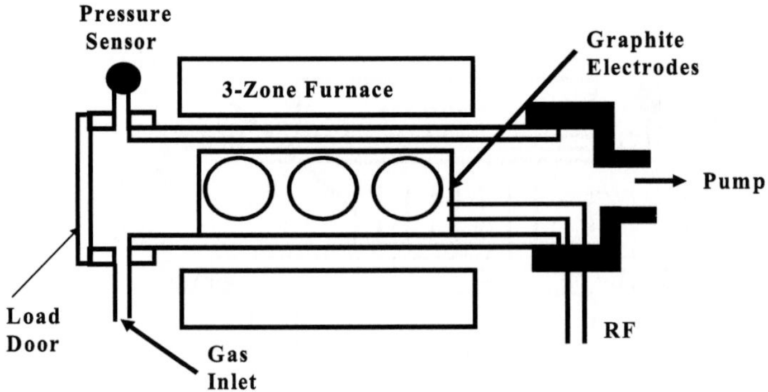

Figure 1-10. Sketch of hot-wall plasma deposition reactor. (*After* [24].)

Then, connection of the alternate slabs to a power supply, induces a glow discharge of the gas flowing in the space between electrodes, which runs parallel to the wafers. By taking the energy for the reaction from the glow discharge, the deposition may be achieved at a wafer temperature in the range of 100 to 350 °C, e.g., Table 1-4.

Table 1-4. Common plasma-assisted CVD reactions for depositing pertinent materials [24].

Product	Reactants	Deposition temperature (°C)
Plasma silicon dioxide	SiH_4+N_2O	200-350
		200-350
Plasma silicon nitride	SiH_4+NH_3	200-350
	SiH_4+N_2	200-350

1.2.1.5 Sputtering

While deposition via CVD requires high temperatures to facilitate gas dissociation, and migration once the atoms/molecules reach the wafer surface, sputtering involves a totally different mechanism. In sputtering, a plasma is created by ionizing an inert gas, typically Argon, at low pressures, e.g., ~10mTorr. The material one wants to deposit on the wafer originates in the bombardment with high energy (typically Argon, Ar^+) ions, present in the plasma above the target substrate containing the material to be deposited on the wafer. Target (cathode) bombardment causes the ejection, via momentum transfer, of its surface atoms, Fig. 1-11. The ejected atoms, in turn, fly off from the target and come to rest on other surfaces within the chamber, in particular, the wafers of interest. The material transfer process is atomic in nature, therefore, its transfers to the wafer in the same ratio it present in the target.

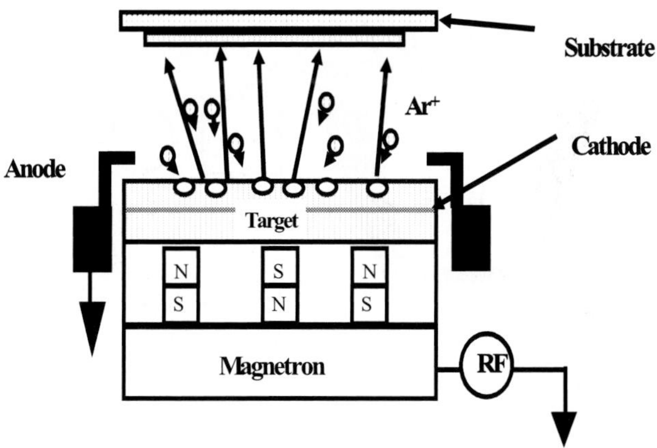

Figure 1-11. Sketch of sputtering deposition system.

Magnetron sputtering is one of the most versatile sputtering techniques because it can be employed to deposit both insulating and non-insulating materials, e.g., Ti, Pt, Au, Mo, W, Ni, Co, Al_2O_3, SiO_2, Fe, Cr, Cu, FeNi, TiNi, AlN, SiN, etc. The technique is based on creating a plasma by inducing the breakdown of an inert gas, e.g., Ar, in the presence of a strong magnetic field. The resulting Ar+ ions are accelerated by the potential gradient between cathode and anode, impinge on the target and, thus, create the flux of material towards the substrate to be coated. Typical maximum thickness of deposited materials is $\sim 5\mu m$.

1.2.1.6 Evaporation

In this deposition technique, the evaporant, the material one wants to deposit on the wafer, is heated off a crucible. Heating may be effected by resistive means or by direct electron-beam bombardment, Fig. 1-12. In the resistive heating approach, the wafers to be coated and the crucible containing the evaporant, are placed inside a vacuum chamber and the latter heated until its vapor pressure is greater than that originally existing in the chamber. Evaporation results in coating everything inside the chamber, in particular, the wafers of interest. In the electron-beam bombardment approach, line-of-sight coating is obtained.

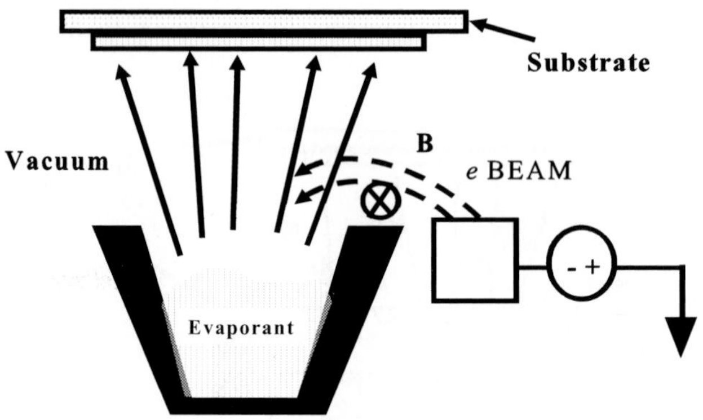

Figure 1-12. Sketch of electron-beam-based evaporation system.

Typical materials deposited by this technique include Al, Cr, Au, Ni, Fe, Ti, Cu, Pt, FeNi, TiNi, SiW, MgO, SiO_2, Al_2O_3, AlN, SiN. The deposition rate is a function of the distance between the evaporant and the substrate, and its typical maximum thickness is usually $\sim 5\mu m$.

1.2.2 MEMS Fabrication Methods

The creation of moveable structures necessitates extending the 2-D IC fabrication process to include shaping of the third dimension, perpendicular to the substrate; this is exemplified, in silicon, by four fundamental techniques, namely Surface Micromachining, Bulk Micromachining, Deep Reactive Ion Etching (DRIE), and single crystal silicon reactive etch and metal (SCREAM), which are presented next.

1.2.2.1 Surface Micromachining

In surface micromachining, 3-D mechanical structures are constructed in a layered fashion. Two types of layers, based on their material composition/etching properties, are employed, namely, sacrificial and structural layers. The former are ultimately dissolved via a process step named *release*, and the latter remain, becoming part of the free-standing movable structure proper. The simplest element illustrating the surface micromachining technique is, perhaps, the cantilever beam. Figure 1.13 sketches its formation. Typical combinations of sacrificial and structural materials, and corresponding etchant are shown in Table 1.5 [27].

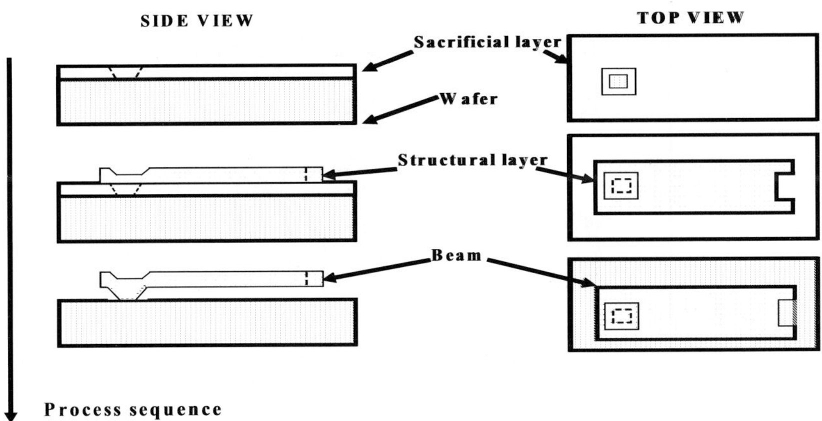

Figure 1-13. Sketch of the formation of a cantilever beam by surface micromachining. From top to bottom of the figure, the sacrificial material is deposited and patterned (top), then the structural material is deposited and patterned (middle), the sacrificial layer is released (bottom).

Table 1-5. Structural/Sacrificial/Etchant Material Systems [27].

Structural Material	Sacrificial Material	Etchant
Aluminum	Single-crystal silicon	EDP, TMAH, XeF_2
Aluminum	Photoresist	Oxygen plasma
Copper or Nickel	Chrome	HF
Polyimide	Aluminum	Al etch (Phosphoric, Acetic, Nitric Acid)
Polysilicon	Silicon dioxide	HF
Photoresist	Aluminum	Al etch (Phosphoric, Acetic, Nitric Acid)
Silicon dioxide	Polysilicon	XeF_2
Silicon nitride pr Boron-doped polysilicon	Undoped polysilicon	KOH or TMAH

1.2.2.2 Bulk Micromachining

As the name implies, bulk micromachining sculpts the substrate itself to form the 3-D mechanical structure. The simplest example of this technique is illustrated by the creation of a cavity, shown in Figure 1.14. As suggested, the aspect ratio of the cavity or pit is determined by the etching properties of the atomic planes which, in turn, are function of the crystallographic properties and orientation of the wafer, in particular, the greater the number of atoms on a given plane, the slower its etching rate. To understand this statement we explain the concept of Miller indices [28].

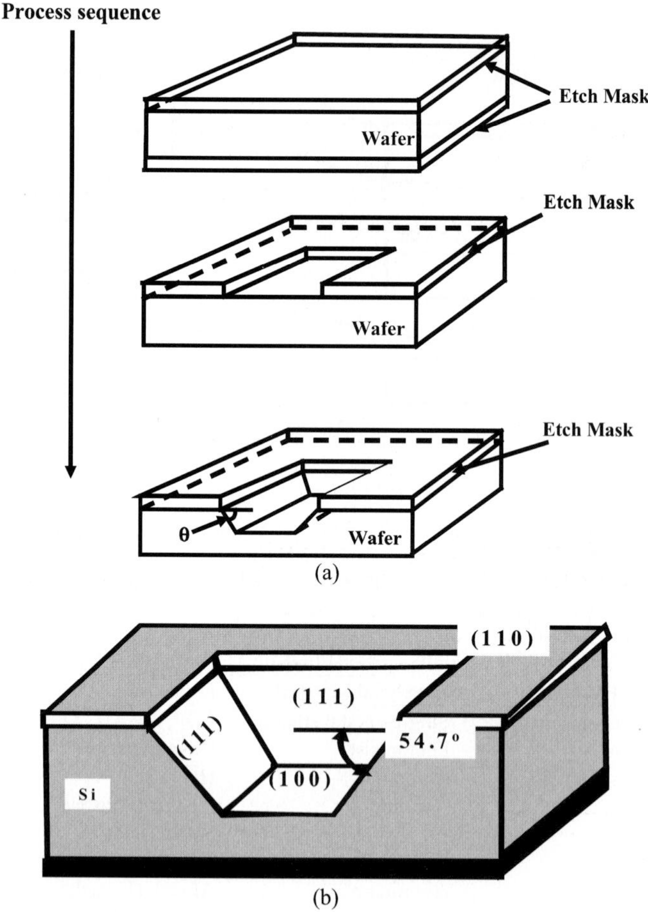

Figure 1-14. Sketch of bulk micromachined cavity. (a) From top to bottom of the figure, a mask is deposited (top), then patterned to expose the wafer (middle), and then the wafer is exposed to an etchant (bottom). (b) Cavity walls are delimited by the crystallographic planes of the wafer.

The arrangement and orientation of atoms in a crystalline solid is specified with reference to certain directions, see Figure 1-15. Thus, with respect to the origin of a Cartesian set of coordinates, the position of an atom may be described as being

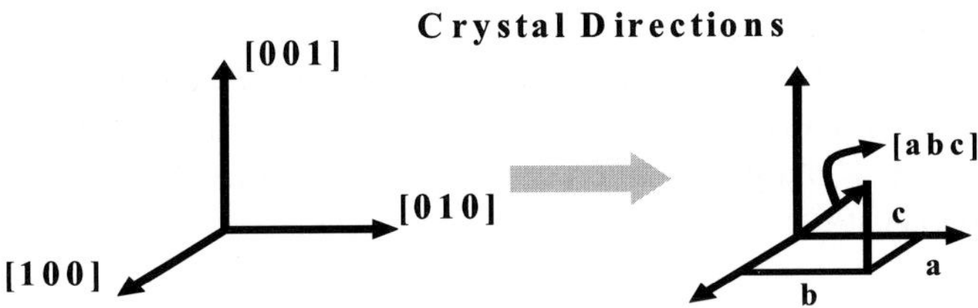

Figure 1-15 Nomenclature of crystal directions.

[abc], that is, a units along the direction [001], b units along the direction [010], and c units along the direction [001]. Since a plane may be described by a vector perpendicular to it (its normal), the direction [abc] also describes a plane, which is denoted the plane (abc), shown in Figure 1-16(a).

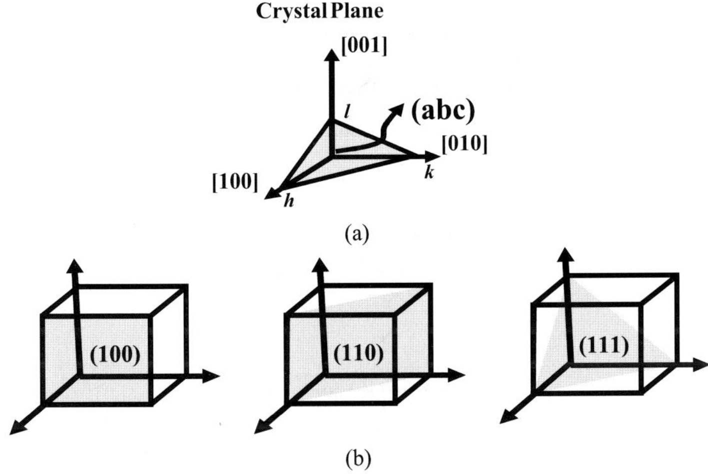

Figure 1-16. (a) Description of crystallographic plane by its normal (abc). (b) Description of crystallographic planes of cubic (atoms occupy the corners and faces of a cube) crystal by Miller indices.

Notice that, since a plane is described by three points common to it, the points of intersection between a plane and the three coordinate axes may also be used to denote it. In particular, see Figure 1-16(b), the points *h*, *l*, and *k*, along the coordinate axes [100], [010], and [001], respectively, might be used for this purpose. However, to accommodate the possibility that the plane might be parallel to one of the coordinate axes, in which case the intersection would occur at infinity, the reciprocals of these points of intersection, (*1/h*, *1/l*, *1/k*), are used instead. Figure 1-16(b) shows examples crystallographic planes and their corresponding of Miller indices [28] for a cubic crystal such as silicon.

The fact that the aspect ratio of bulk micromachined structures is limited by the natural inclination of the crystallographic planes making up the walls, motivated the development of techniques to increase it. The sections below address two of these.

1.2.2.3 Deep Reactive Ion Etching

The idea behind DRIE is to achieve high-aspect ratio trenches by selectively enhancing the etch rate at the bottom of the trench, while inhibiting the lateral etch rate. This is accomplished by combining a sequence of plasma etching and polymerization steps [31], [32], see Figure 1-17(a).

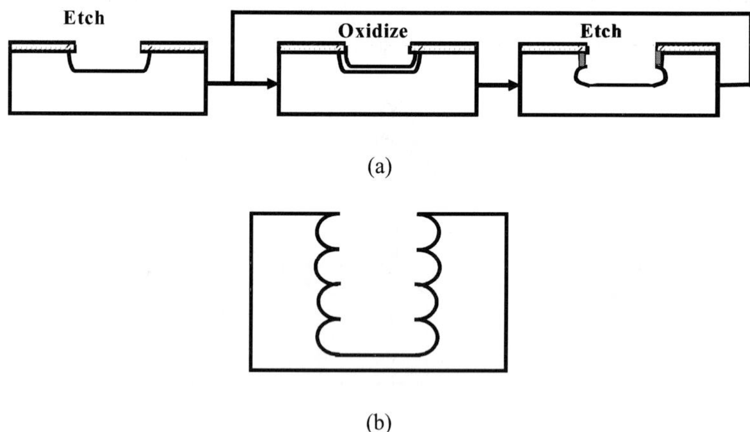

Figure 1-17. Deep reactive ion etching (a) Etching/polymerization sequence. (b) Wall scalloping.

During the plasma etching steps, as indicated previously, positive ions resulting from the breakdown discharge of a gas above the silicon wafer, bombard the silicon surface as they fall vertically towards the negatively charged wafer. To achieve vertical selectivity, the sidewalls are protected by a polymer (PR). Thus, this results in etching being primarily effected at the bottom of the trench. Each etching step, which may result in a lateral etch of tenths of microns, is stopped after the maximum tolerated lateral etch is produced. By repeating the passivation/etch sequence, trenches with overall depths of up to several hundreds of microns have been demonstrated. The process proceeds at room temperature, can produce selectivities of 200:1 in standard PR masks, 300:1 in hard masks such as SiO_2 and Si_3N_4, and exhibits etching rates of $6\mu m / sec$ [30]. As a result of this process, the walls of the etched trenches exhibit a scalloping structure, see Figure 1-17(b). The application of DRIE requires acquiring the DRIE equipment. An alternative to DRIE for better than conventional bulk micromachining, but not as expensive as DRIE, is presented next.

1.2.2.4 Single Crystal Silicon Reactive Etch and Metal (SCREAM)

Similar to DRIE, the single crystal silicon reactive etch and metal (SCREAM I) process effects bulk micromachining using plasma and reactive ion etching (RIE) [33], see Fig. 1.18. The process, however, employs standard tools, is self-aligned, employs one mask to define structural elements and metal contacts, and employs a temperature below 300 °C. This low temperature capability makes it amenable for integration of MEMS devices with very large scale integration (VLSI) technology [33].

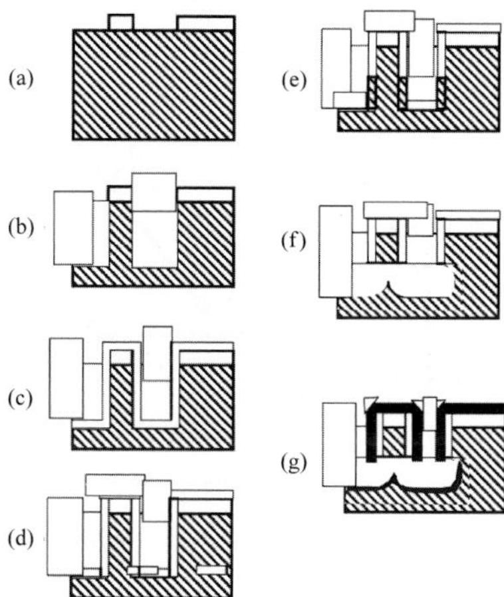

Figure 1-18. SCREAM I process flow. (a) Deposition and patterning of PECVD masking oxide. (b) RIE of silicon with BCl_3/Cl_2. Typically 4-20 μm deep. (c) Deposition of oxide sidewall via PECVD, typically 0.3 μm thick. (d) Vertical etch of bottom oxide with CF_4/O_2 RIE. (e) Etch of silicon 3-5 μm beyond end of sidewall with Cl_2 RIE. (f) Isotropic RIE release of structures with SF_6 RIE. (g) Sputtering deposition of aluminum metal. The device shown is a beam, free to move left-right, and its corresponding parallel-plate capacitor. (*After* [33].)

1.2.3 Nanoetechnology Fabrication Elements

The elements of nanotechnology fabrication range from techniques to produce two-dimensional patterns with deep-submicron/nanometer-scale widths, to techniques to produce atomic-thick layers/multi-layers of various material compositions, to techniques to precisely manipulate atomic-size particles. These techniques, together with those presented previously, constitute the arsenal at the core of NanoMEMS.

1.2.3.1 Electron Beam Lithography

Electron beam lithography utilizes electrons, instead of the projection of a mask image illuminated by photons, to create directly the desired pattern on the PR, Figure 1-19.

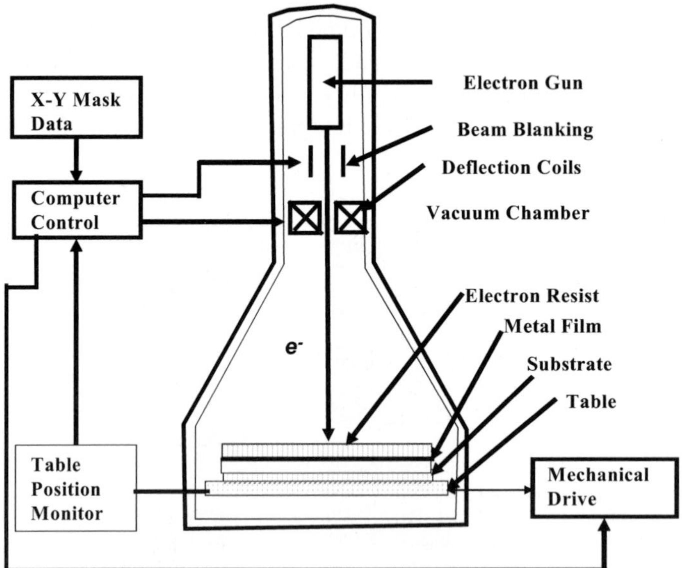

Figure 1-19. Sketch of electron bean lithography system. (*After* [23].)

Since the wavelength of electron accelerated through a potential difference V is $\lambda(\AA) = \sqrt{150/V}$, an electron beam may be focused to a diameter of $0.01 - 0.5\mu m$, and resolutions of 1nm are obtained. The electron beam is focused and scanned either in a raster (sequential) fashion, or in a vector fashion where the image field consists of independently addressable/exposable pixels, Fig. 1-20.

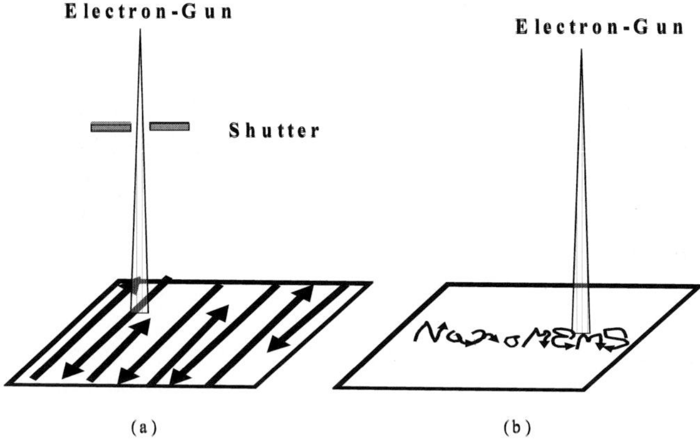

Figure 1-20. Electron-beam patterns. (a) Raster scan. (b) Vector scan.

The ultimate resolution of electron-beam lithography is not posed by beam spot size, but by the so-called electron scattering and proximity effects, Figures 1-21, 1-22.

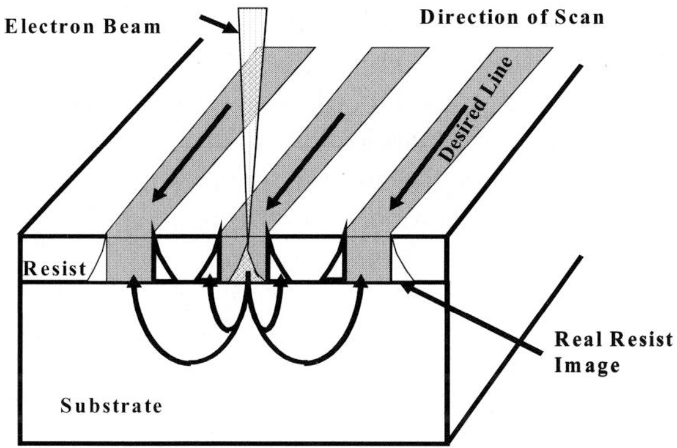

Figure 1-21. Sketch of electron scattering effects on PR-coated wafer substrate. (*After* [23].)

The former captures the fact that, in the course of penetrating the PR and underlying substrate, the electron beam scatters and experiences a directional change manifested as a spreading out of the beam, i.e., increase in its spot size. The latter, in turn, captures the fact that some of the scattered electrons are absorbed, not under the profile of the beam spot, but in areas adjacent to it. Two more effects resulting from beam scattering produce width- and proximity-dependent patterns, Figure 1-22.

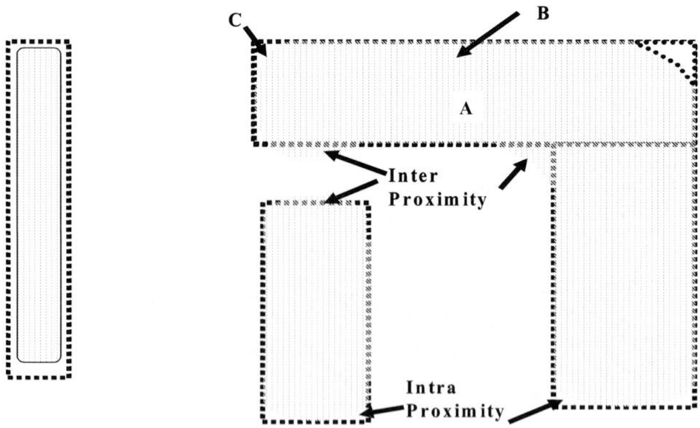

Figure 1-22. Intra- and inter-proximity effects due to electron scattering. (*After* [23].)

The intra-proximity effect reflects the fact that the PR area near the center of the beam spot receives more energy, from adjacent points, than the PR nearest to the circumference. Thus corners, like point *A*, tend to be underexposed. The inter-proximity effect, on the other hand, reflects the fact that electrons intended to define one pattern scatter unto adjacent patterns, thus extending the effective width of the adjacent pattern. Reflecting all these factors, the highest resolution of electron beam lithography as employed for nanoscale device fabrication is about 10nm, however, the slow nature of writing the patterns one at a time, makes this technique expensive and not amenable for mass production. Its main applications are in the creation of masks and in nanotechnology research.

1.2.3.2 Soft Lithography

The conventional IC fabrication processes, and the approaches to MEMS fabrication derived from them, have as their core step the photolithographic definition of patterns on a planar substrate/wafer. Thus, as indicated previously, their application to creating nanoscale devices becomes prohibitively expensive, as the development of the concomitant light sources and tools to create devices at these length scales is very expensive. This is of chief import, not just for research purposes but, more importantly, for the large scale production germane to commercial applications.

Soft lithography, the production of nanoscale devices by creating elastic (soft) polymer masters that can then be used to print, mold, and emboss nanoscale structures, is a technique which has been the subject of much recent research for the inexpensive creation of nanoscale devices. The technique relies on first making an elastic stamp, shown in Figure 1-23, and

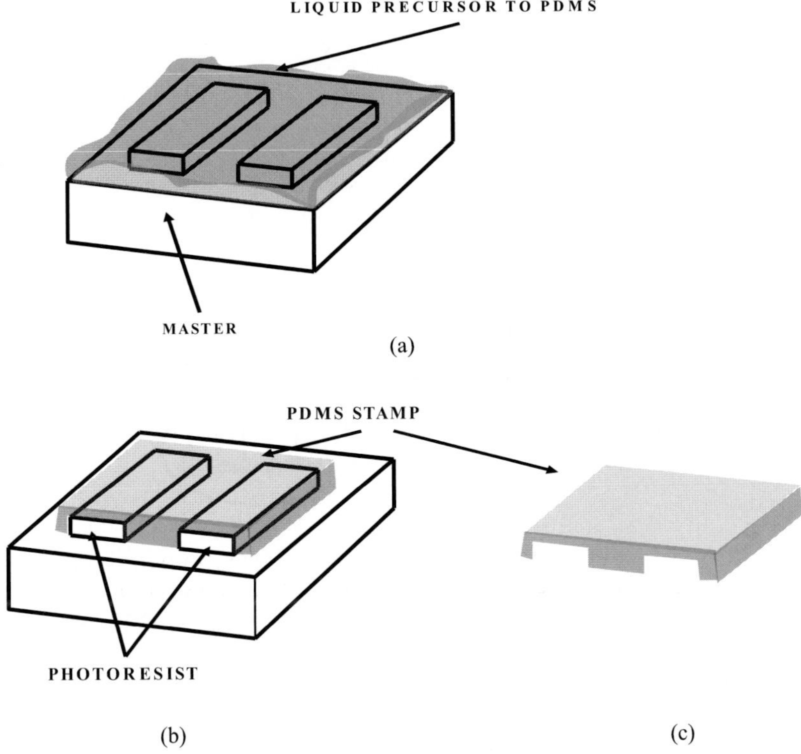

Figure 1-23. Soft lithography—Making an elastic stamp. (a) A liquid precursor to polydimethylsiloxane (PDMS) is poured over a bas-relief master produced by photolithography or electron-beam lithography. (b) The liquid is cured into a rubbery solid that matches the original pattern. (c) The PDMS stamp is peeled off the master. (*After* [34].)

appears to have been advanced by Whitesides [34], who applied it as an extension of his work on the creation of channels and chambers for microfluidic systems.

Printing is effected by inking the elastic stamp with a solution of organic molecules called thiols, and pressing it against a thin film of gold that has been deposited on a silicon wafer, Figure 1-24(a). Due to the nature of the chemical interaction between the thiol molecules and the gold, the surface is wetted with the thiols displaying a preferred orientation and creating a self-assembled monolayer, Figure 1-24(b), which delineates the stamp's pattern. The feature size or minimum width of the pattern is of the order of 50nm [34].

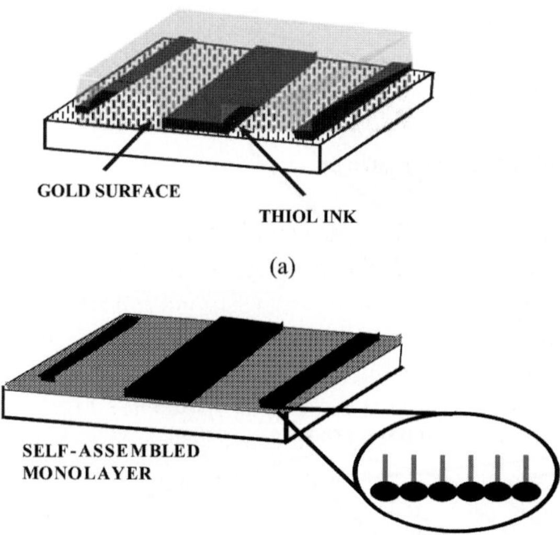

Figure 1-24. Microcontact printing. (a) The elastic stamp (PDMS) is inked in thiols and then pressed against the gold film previously deposited in the wafer. (b) The stamp is retracted, transferring a pattern of self-assembled thiols. *(After* [34].)

Molding is effected by pressing the elastic stamp against a liquid polymer on the wafer, shown in Figure 1-25, which causes the polymer to flow into

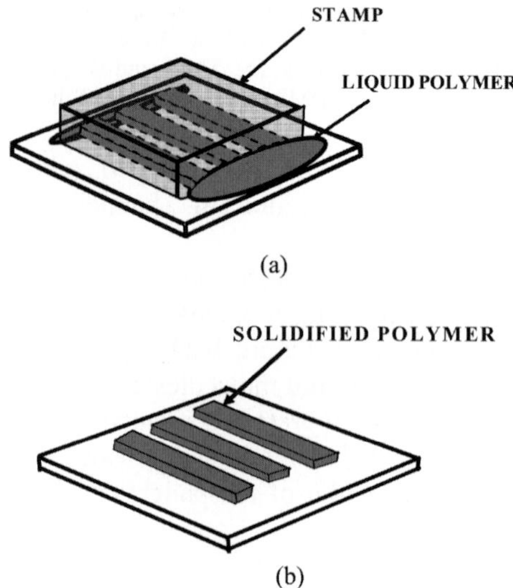

Figure 1-25. Molding. (a) The elastic stamp is pressed against the deposited liquid polymer, which flows into the recesses/channels of the mold. (b) Upon curing, the polymer solidifies into the mold pattern. *(After* [34].)

the stamp's recesses. Then, upon curing the polymer, this solidifies according to the stamp's pattern. The feature size for patterns thus created may be as small as 10 nm [34].

1.2.3.3 Molecular Beam Epitaxy

The engineering of modern semiconductor device structures relies on the appropriate introduction and distribution of impurities via doping, together with band-gap engineering to effect electron confinement along the direction of transport [34-37]. This latter gives rise to devices in which tunneling phenomena becomes manifest. The key to these types of structures is the technique for depositing down to mono-atomic-thick layers called molecular beam epitaxy (MBE). MBE underwent extensive progress during the 1990s and is now a well established production technology [38].

The essentials of MBE for growing a given structure are depicted in Figure 1-26.

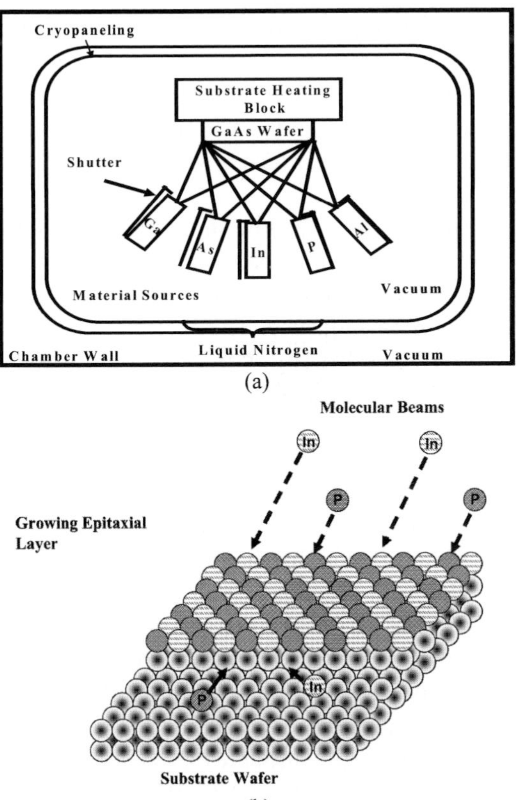

Figure 1-26. (a) Sketch of MBE system. The atomic sources may be either in the solid or the gaseous states. (b) Sketch of layered atomic deposition. (*After* [38].)

The system consists of a steel chamber which is equipped with pumps, to create a very low pressure environment, typically about 10^{-11} Torr, and a growth chamber containing several vacuum furnaces, called effusion cells or K-cells, from where a variety of atomic or molecular materials evaporate. The target wafer, on which growth is to occur, is placed inside the chamber where it is held at a high, controlled temperature and under high vacuum, and rotated to achieve uniformity over the wafer.

Growth occurs when heating of the K-cells causes the various materials in them to evaporate, thus forming atomic beams that land on the wafer surface. The properties of the growing layers are controlled by a number of parameters, particularly, K-cell temperature, which controls beam intensity or atomic/molecular flux, and substrate temperature, which controls the dynamics of the atoms once these reach the wafer surface, see Figure 1-26(b). In particular, the arriving atoms evolve according to the following competing mechanisms: 1) Immediate absorption to the surface, i.e., they "stick" wherever they land; 2) Migration across the surface, i.e., move around before coming to a resting place which may not preserve the crystalline structure; 3) Incorporation into the crystal lattice; and 4) Thermal desorption, i.e., they reevaporate from the surface. To achieve good crystal quality, such a set of flux and substrate temperature parameters must be discovered that the arriving atoms have sufficient energy to move to the appropriate position on the surface, without re-evaporating, and be incorporated on the crystal.

The MBE technique is very versatile in that it allows the composition of the layers to be fine tuned. This is accomplished by equipping the K-cells with shutters which, through computer control, can turn on or off each beam according to precise timing sequence. The fact that growth is controlled by computer, endows MBE with the ability to deliver even atom-thick layers, of abrupt composition, in a reproducible and reliable fashion. This, in turn, enables bandgap engineering, the use of the material band gap as a degree of freedom to engineer device properties. In the InP HBT, an emitter with a band gap greater than that of the base, permits high base doping, without compromising current gain, by virtue of the reduction of hole current injection into the emitter effected by the latter's energy barrier. In the RTD, a lower band gap region, a potential well, sandwiched between two large band gap regions, barriers, allows preferential current conduction only when the energy of conduction electrons coincides with the discrete energy state in the potential well, thus giving rise to the creation of a current-voltage characteristic exhibiting negative differential resistance. The fact that the path length of electron transport through the device is very short, leads to these devices exhibiting very high speeds of operation, e.g., hundreds of GHz in the case of the HBT, and close to a THz in the case of the RTD.

Figures 1-27(a) and (b) show the layer structures of MBE-grown heterostructure bipolar transistor (HBT) and resonant tunneling diodes (RTD), respectively.

Thickness	Layer	Doping
100 nm	GaInAs Contact	n=1x10^{19}cm^{-3}
70 nm	AlInAs Emitter Contact	n=1x10^{19}
120 nm	AlInAs Emitter	n=8x10^{17}
30 nm	Compositional Grade	n=8x10^{17}
10 nm	GaInAs Spacer	p=2x10^{18}
60 nm	GaInAs Base	p=2x10^{19}
20 nm	GaInAs Spacer	p=5x10^{17}
50 nm	GaInAs Spacer	n=1x10^{17}
750 nm	InP Collector	n=3x10^{16}
700 nm	GaInAs Subcollector	n=1x10^{19}
10 nm	GaInAs Buffer	Undoped
	InP Substrate	

(a)

Layer	Thickness
GaInAs contact (n+=5E18)	2000Å
GaInAs spacer (n=5E17)	250Å
GaInAs spacer (und.)	15Å
AlAs barrier	13Å
GaInAs/InAs/GaInAs well	12Å/30Å/12Å
AlAs barrier	13Å
GaInAs spacer (und.)	15Å
GaInAs spacer (n=1E17)	250Å
GaInAs contact layer (n+=5E18)	5000Å
GaInAs buffer (und.)	100Å
InP substrate (semi-insulating)	

(b)

Figure 1-27. Layer description of MBE-grown devices. (a) InP double heterostructure bipolar transistor (DHBT) [39]. (b) Resonant tunneling diode (RTD) [40].

1.2.3.4 Scanning Probe Microscopy

Progress in Nanotechnology has been intimately related to the invention of a number of techniques for imaging and manipulating atoms/nanoparticles at nanoscales. All of these techniques are based on a very fine tip (with atomic resolution), and the nature of what is imaged or manipulated is a function of the tip itself, i.e., whether it is conductive, insulating, magnetic, non-magnetic, etc. Excellent review articles summarizing advances in scanning probe microscopy has been published recently by Giessibl [41] and Baski [42]. In this section we focus on two of the main such techniques, namely: 1) The scanning tunneling microscope (STM); 2) The atomic force microscope (AFM).

1.2.3.4.1 Scanning Tunneling Microscope

In STM, a sharp metal tip is brought in very close proximity to a conductive sample, typically to a distance within a few Angstroms, see Figure 1-28 [16].

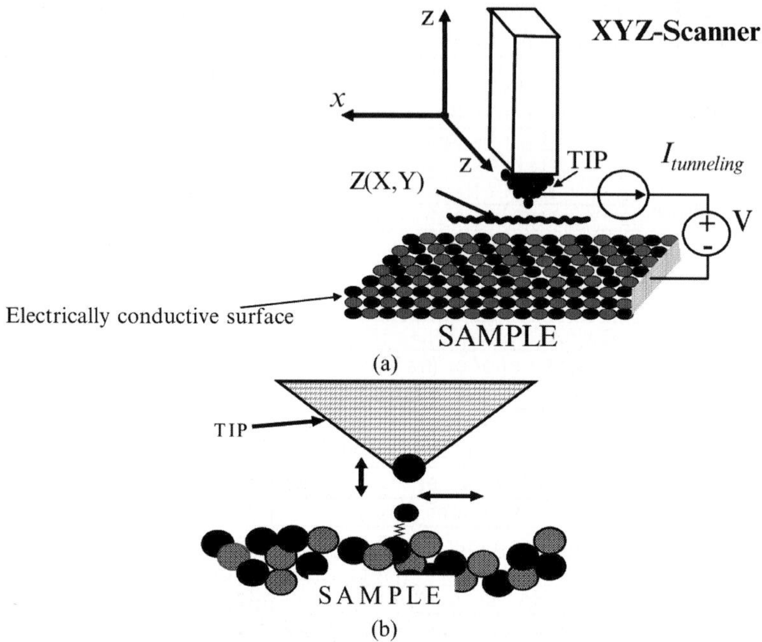

Figure 1-28. (a) Sketch of STM system. (b) Probe tip detail. The sample is held in ultra high vacuum. (*After* [16].)

Then, when a bias voltage is applied between the tip and the conductive sample, electrons tunnel quantum mechanically across the air gap to elicit a tunneling current of a magnitude not exceeding several nA. Due to the nature of the tunneling current I_t, which obeys the equation $I_t(z) = I_0 e^{-2\kappa_z z}$, where $\kappa_z = \sqrt{2m\Phi}/\hbar$ embodies the properties of the tunneling electron (its mass *m*), and the work function of the tip material Φ, with $\hbar$ being Planck's constant, the tunneling current is a very sensitive function of the tip-sample distance, z. Imaging, therefore, may be produced in two modes: 1) Scanning the tip in the x-y plane while forcing it to remain at a fix z-position. This, so called *constant height mode*, extracts sample morphology/relief image from modulation of the tunneling current magnitude as the variations in the sample relief change the tip-sample distance. Thus, an image of $I_t(x, y, z \approx constant)$ is obtained; 2) Scanning the tip in the x-y plane while adjusting the tip position z to keep the tunneling current constant. This is the called *topography mode*, and produces an image of $z(x, y, I_t \approx constant)$.

STM tips are fabricated via chemical etching or mechanical grinding of W, Pt-Ir, or pure Ir [41]. By using a magnetic probe tip the STM can be made sensitive to the spin of the tunneling electrons. Besides the tip sharpness and material properties, moving the tip with atomic scale precision, to obtain atomic resolution image, necessitates the utilization of a piezoelectric ceramic, whose extremely fine deformation is induced by an applied voltage.

1.2.3.4.2 Atomic Force Microscopy

In AFM, Figure 1-29, a sharp tip is also brought very close to the sample surface.

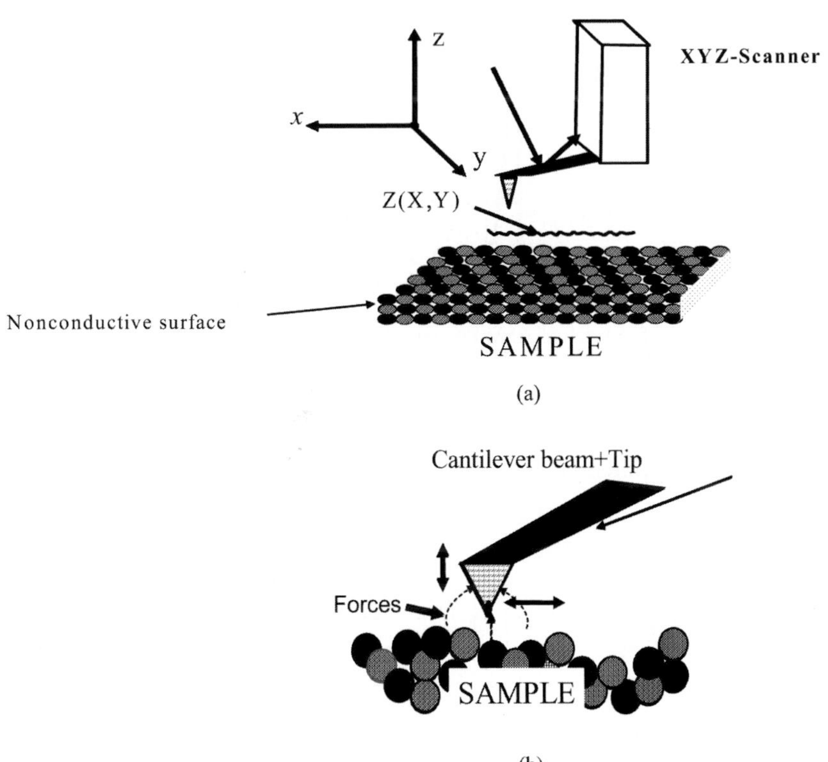

Figure 1-29. (a) Sketch of AFM system. (b) Probe detail. The sample may be held at ambient conditions. (*After* [41].)

However, unlike STM, no voltage is applied between the tip and the sample. Instead of a tunneling current, the AFM detects the force elicited between the tip and the sample. The tip is part of a force-sensing cantilever beam so that, when the latter is raster-scanned over the sample, much like a phonograph, surface height variations are detected by monitoring the interference pattern produced by a laser beam reflecting off the cantilever beam when the latter deflects/deforms.

The image of the sample is then extracted by relating the cantilever beam deflection to the force required to produce it, F_{TS}. F_{TS} in turn, is related to the tip-sample (TS) potential V_{TS} via its negative gradient, $F_{TS} = -\partial V_{TS} / \partial z$ and is characterized by an effective spring constant $k_{TS} = -\partial F_{TS} / \partial z$. F_{TS} may be attractive or repulsive, as it embodies a variety of forces, each one varying differently with TS distance z, thus making it a nonlinear force, see Figure 1-30.

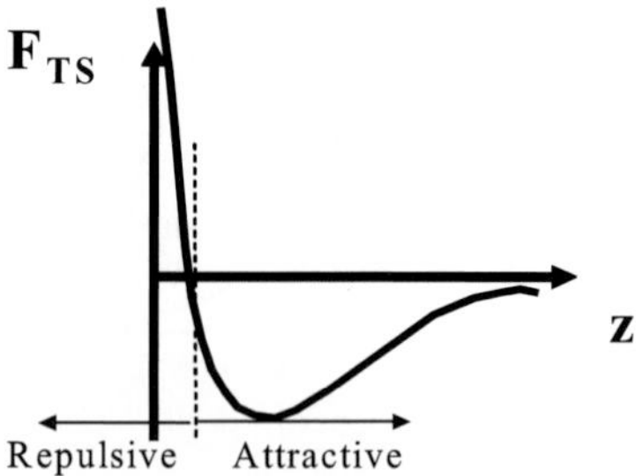

Figure 1-30. Sketch of AFM tip-sample force versus their separation z.

For instance, at distances under 1nm, short-range chemical forces are operative which, for anisotropic chemical bonds, are best characterized by a Stillinger-Weber potential, $V_{SW} = V_n + V_{nn}$ where both nearest neighbor potential V_n, given in Eq. (1), and next nearest

$$V_n(r) = E_{bond} A\left[B\left(\frac{r}{\sigma'}\right)^{-p} - \left(\frac{r}{\sigma'}\right)^{-q}\right] e^{\frac{1}{r/\sigma' - a}} \quad for\ r < a\sigma',\ else\ V_{nn}(r) = 0 \qquad (1)$$

potential V_{nn} given in Eq. (2), and (3) are considered.

$$V_{nn}(r_i, r_j, r_k) = E_{bond}\left[h(r_{ij}, r_{ik}, \theta_{ijk}) + h(r_{ji}, r_{jk}, \theta_{ijk}) + h(r_{ki}, r_{kj}, \theta_{ikj})\right], \qquad (2)$$

with

$$h(r_{ij}, r_{ik}, \theta_{jik}) = \lambda e^{\gamma\left(\frac{1}{r_{ij}/\sigma' - a} + \frac{1}{r_{ik}/\sigma' - a}\right)} \left(\cos\theta_{jik} + \frac{1}{3}\right)^2 \quad for\ r_{ij,ik} < a\sigma',\ else\ 0. \quad (3)$$

The optimal parameters, in terms of experimental agreement for a silicon tip on a silicon sample, was found by Stillinger and Weber to be as follows: A = 7.049556277, p = 4, γ = 1.20, B = 0.6022245584, q = 0, λ = 21.0, E_{bound} = 3.4723 aJ, a = 1.8, σ' = 2.0951 Å, and $\sigma = 2^{1/6}\sigma'$.

Similarly, at distances under 100nm, long-range forces, namely, van der Waals, electrostatic, and magnetic forces are operative. The van der Waals forces, are characterized by a potential given by Eq. (4)

$$V_{vdW} = -\frac{\alpha d_1^2}{z^6}. \quad (4)$$

For the tip-sample situation found in AFM, namely, a spherical tip with radius R separated a distance z from a flat surface (where z is the effective distance between the plane connecting the centers of the surface atoms and the center of the closest tip atom) the van der Waals potential is given by [42] Eq. (5)

$$V_{vdW} = -\frac{HR}{6z}, \quad (5)$$

where H is the Hamaker constant embodying the atomic polarizability and density of the tip and sample material pair and, for the majority of solids and interactions across vacuum, has a value of $H = 1eV$. For tip-sample materials characterized by this value of Hamaker constant, and with a spherical tip of radius R~100nm separated from flat sample by ~0.5nm, the respective van der Waals potential and force are approximately -30eV and -10nN, respectively.

When both the tip and the sample are conductive and at separations of ~100nm, they may also experience electrostatic forces, characterized by the potential, Eq. (6) [42-45]:

$$F_{electrostatic}(z) = -\frac{\pi\varepsilon_0 RV^2}{z}, \quad (6)$$

where V is the electrostatic potential difference. Accordingly, a potential difference V~1Volt, between a spherical tip of radius R~100nm a distance z~0.5nm from a flat surface, will experience a force ~-5.5nN.

Based on the method employed to extract F_{TS}, and hence the surface image, AFM operation is classified following three modes:

1) Contact Mode-Static AFM: In this mode the tip is in repulsion regime and exerts a large normal and lateral force on the sample. The force applied to the cantilever is kept constant during the scan by applying feedback, while the z-displacement is measured yielding the surface topography. The main drawback of this technique is that it can only be applied in certain cases, namely, at low temperatures, due to the need to circumvent its low-frequency noise and thermal expansion effects on resonance frequency [42].

2) Non-Contact Mode-Dynamic AFM: In this mode the cantilever is mounted on an actuator which vibrates and, thus, excites it with amplitude A_{drive} and frequency f_{drive} to oscillate above the sample. The tip-sample distance is such that operation is in the attractive regime. This may avoid the force and noise problems of contact mode, but is subject to jump-to-contact if the spring constant corresponding to the tip-sample potential overcomes that of the cantilever, i.e., if $k < k_{\max TS}$. The imaging signal is derived from the change in cantilever amplitude and phase that result when the tip approaches the sample. Since the excitation signal may consist of, either fixed amplitude and fixed frequency, or fixed amplitude and varying frequency, these two modes of operation are distinguished. The former is called *AM-AFM* and, while this method does provide atomic resolution, the fact that the time required to capture the tip-surface interaction $\tau_{AM} \approx 2Q/f_0$ is proportional to the quality factor (Q) of the cantilever, which may be tens of thousand, makes it relatively slow.

The latter mode, in which the amplitude is fixed, but the frequency is varied, is called *FM-AFM* mode of operation. This mode also provides atomic resolution, but it is much faster than AM-AFM because the tip-surface interaction time is only $\tau_{FM} \approx 1/f_0$.

3) Intermittent Contact Mode-Dynamic AFM : In this mode the tip is excited to oscillate above sample, also in the attractive regime, but it is made to contact (“tap”) the sample for a short time during every cycle.

One of the key aspects of AFM is the design of the cantilever, particularly, its spring constant and resonance frequency. These are given by Eqs. (7) and (8), respectively, for a beam of thickness t, width w, length L, Young’s modulus E, and mass density ρ.

$$k = \frac{Ewt^3}{4L^3} \tag{7}$$

$$f_0 = 0.162 \frac{t}{L^2} \sqrt{\frac{E}{\rho}} \tag{8}$$

Accordingly, various aspects, which depend on the application, must be considered in designing the cantilever. For example, in the static AFM mode, the spring constant must be chosen so that the beam easily deflects in response to the tip-sample force. Thus, for k_{TS} between 10N/m and 100N/m, the rule is to choose k between 0.01N/m and ~5N/m, with typical resonance frequencies of 2kHz.

On the other hand, for the dynamic AFM techniques it has been found that, to avoid jump-to-contact, the product of the cantilever spring constant and the vibration amplitude must exceed the maximum tip-sample attractive force, i.e., $kA_{response} > F_{TS}^{\max}$. This means that there is a trade-off between cantilever stiffness and excitation drive amplitude. In other words, the spring force pulling the cantilever away from its point of closest proximity to the sample, must overcome the maximum attraction force. A refined criterion to avoid jump-to-contact and which assumes the possibility of a hysteretic $F_{TS}(z)$ relationship is given by [45]:

$$\frac{1}{2} kA^2 > \Delta E_{TS} \frac{Q}{2\pi} \tag{9}$$

where ΔE_{TS} is the hysteresis energy supplied to the cantilever beam in each vibration cycle. A typical set of k, A values for FM-AFM are

$$k = 17N/m, \quad A = 34nm.$$

Typically, the AFM cantilevers are fabricated via Si or Quartz micromachining, and the usual tip materials include Si integrated with beam, W, Diamond, Fe, Co, Sm, CoSm permanent magnets, and Ir.

1.2.3.5 Carbon Nanotubes

Carbon nanotubes are, perhaps, the quintessential element of nanotechnology. Their discovery is the fruit of research, originally conducted by Kroto and Smalley in 1985, with the aim of studying the laser vaporization of graphite. Such studies elicited the discovery by them of clusters containing 60 carbon atoms (C_{60}: Buckminsterfullerene), arranged in a spherical structure, see Figure 1-31, [1].

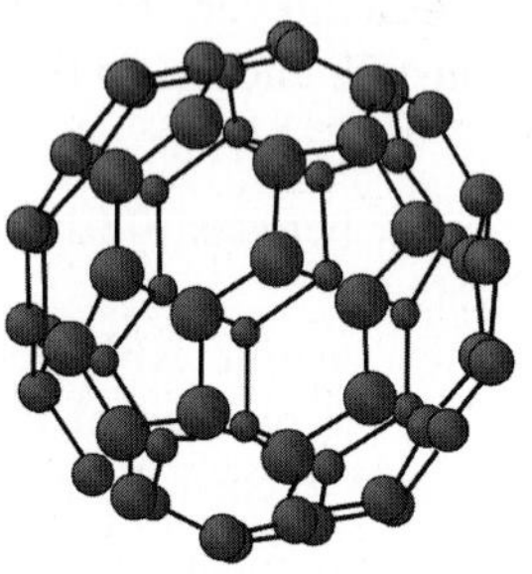

Figure 1-31. Sketch of the chemical structure of C_{60}: Buckminsterfullerene. (*After* [46].)

Continued research to increase the yield of these C_{60} clusters led Iijima to discover carbon nanotubes (CNT), see Figure 1-33 [46].

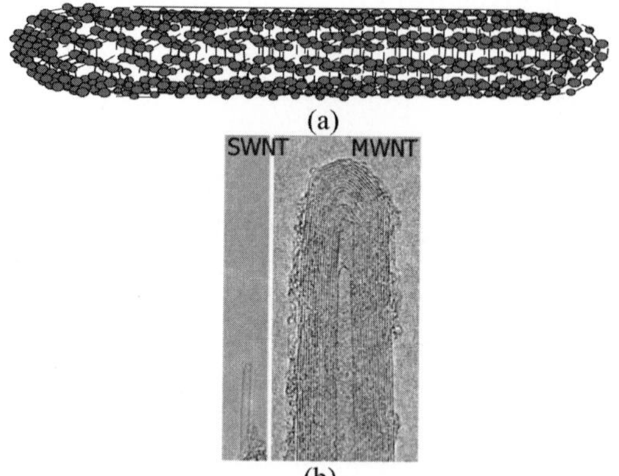

Figure 1-32. (a) Sketch of the chemical structure of a single-wall carbon nanotube (SWNT). (*After* [1].) (b) SEM of SWNT and MWNT. In a multi-walled nanotube, an inner SWNT forms the core of multiple concentric nanotubes which grow around it. (Courtesy of Prof. László Forró, Swiss Federal Institute of Technology (EPFL), Lausanne Switzerland).

CNTs are molecular carbon fibers that consist of graphite cylinders closed at each end by caps containing six pentagonal rings, i.e., each cap is exactly one-half of a C_{60} molecular cluster [46]. They tend to be produced in three main modalities, namely, single-walled nanotubes (SWNTs), which range in diameter from approximately 0.4nm to more than 3nm, multi-walled nanotubes (MWNTs), which range in diameter from approximately 1.4nm to more than 100nm, and ropes, which are parallel stripes of SWNTs stuck to each other. Their physical properties are astounding. With aspect ratios of the order of 10-1000, they are several μm (ropes up to cm) long, possess a Young's modulus, tensile strength, and density of ~1TPa (Steel: 0.2TPa), 45GPa (Steel: 2 GPa), and $1.33 \sim 1.4\, g/cm^3$ (Al: 2.7 g/cm^3). In addition, their conductivity may be metallic or semiconducting, and they have a current carrying capability of $\sim 1\, TA/cm^3$ (Cu: $1\, GA/cm^3$). A number of techniques are employed to produce CNTs, for instance, the arc discharge, laser ablation and chemical vapor deposition methods. These methods usually yield a random mixture of SWNTs, MWNTs, and ropes and research is under way to determine techniques for the controlled growth of a specific type of CNT. For instance, Li *et al.* [47] have reported the development of a catalyst-based method that predominantly yields SWNT. In this method, a silicon wafer is pre-patterned with alumina nanoparticles, which serve as catalysts for their CVD growth, producing SWNTs with diameter under 1.5nm.

The narrow diameter of CNTs makes them ideal candidates for applications as SPM tips, as well as a number of devices, such as channels for field effect transistors. Figure 1-33 shows the formation of CNT tips.

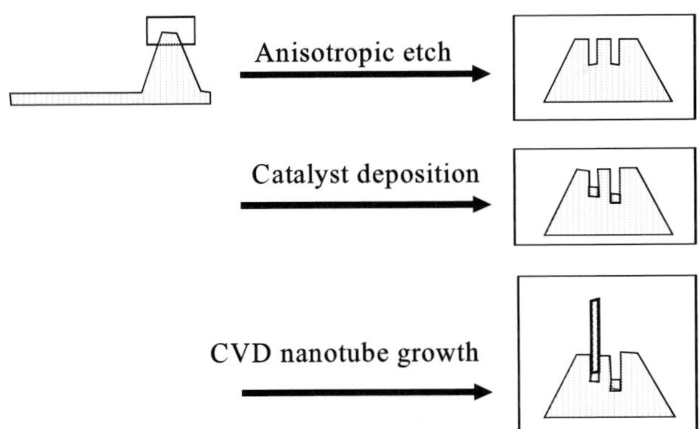

Figure 1-33. Formation of AFM tips via CNT growth. (*After* [48].)

1.2.3.6 Nanomanipulation

The ultimate degree of control in nanofabrication, is embodied in the ability to manipulate individual atoms/nanoparticles with precision. This is accomplished by two techniques, namely, exploiting AFM to push particles, and DIP-Pen lithography.

1.2.3.6.1 AFM-based Nanomanipulation

In this technique, an oscillating AFM tip is brought close to a particle until, as a result of jump-to-contact, the oscillation amplitude goes to zero. The AFM approaches the nanoparticle via a fast X-Y scanning oscillation, in a plane perpendicular to the desired pushing direction, z, see Figure 1-34. Once contact of the AFM with the nanoparticle is established, motion proceeds in the z-direction at a slow scan rate.

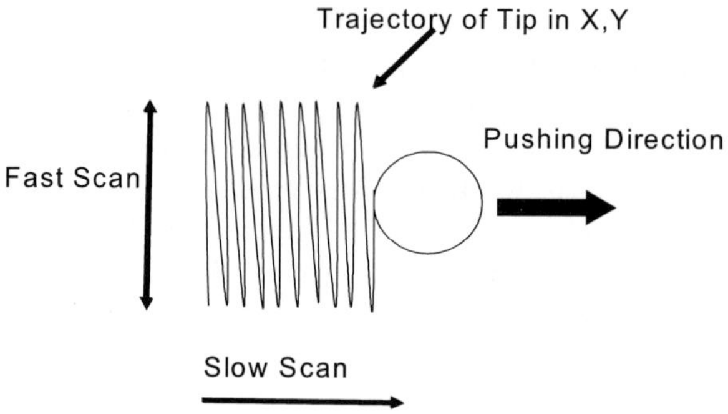

Figure 1-34. Pushing a nanoparticle with AFM. (*After* [49].)

1.2.3.6.2 DIP-Pen Lithography

In this technique, developed by Mirkin's group [50], see Fig. 1-35, and

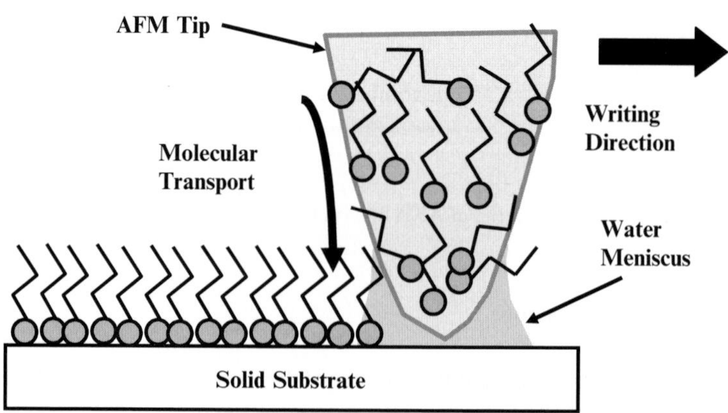

Figure 1-35. Close-up of inked AFM tip as molecules flow down the tip via water meniscus. (*After* [50].)

reminiscent of a goose-feather pen, a molecular “ink” is deposited over a gold surface according to a desired pattern. In one demonstration of the technique, an AFM tip was coated with a thin film of thiol molecules (the “ink”) and moved two-dimensionally so as to inscribe the underlying gold surface.

Since the thiol molecules can only attach to the gold surface in one particular orientation, a self-assembled monolayer of them, embodying the desired “writing,” results. A variety of “inks” may be employed and, in terms of line width capability, lines a few-nanometers wide have been demonstrated.

1.3 Summary

In this chapter we have introduced the broad field of nanoelectromechanical systems. In particular, we have traced its origins, motivation, and presented a unified survey of its distinctive characteristic, namely, the convergence of fabrication techniques, from conventional IC fabrication, to microelectromechanical systems fabrication, to nanoscale fabrication. In the next chapter, we address the fundamental physics on which devices, circuits and systems exploiting the NanoMEMS fabrication methods may be predicated.

[illegible]

1.5 Summary

In this chapter we have introduced the [illegible] mechanochemical systems. In particular, we [illegible] and presented [illegible]

Chapter 2

NANOMEMS PHYSICS: QUANTUM WAVE-PARTICLE PHENOMENA

2.1 Introduction

As discussed in Chapter 1, NanoMEMS aims at exploiting the convergence between nanotechnology and microelectromechanical systems (MEMS) brought about by advances in the ability to fabricate nanometer-scale electronic and mechanical device structures. This novel paradigm, in turn, poses an interesting challenge from the device physics point of view. In particular, the invention and/or discovery of a plethora of new materials, concepts and techniques such as carbon nanotubes (CNTs) [17], photonic band-gap crystals (PBCs) [51], and MEMS [52-55], respectively, has opened up new possibilities to implement novel devices upon which a new "electronics" technology, with attributes that are far superior to everything known to date, may be predicated. With the simultaneous convergence and exploitability, at such small length scales (e.g., down to a few nanometers), of various types of physical properties and effects, for instance, electronic, mechanical, optical, and magnetic and quantum effects, the nature of the concomitant new universe of devices and circuits that will fuel this new electronics will clearly be vast, yet, it is at present mostly unknown. In this context, many domains of physics, not usually invoked in describing the behavior of prior-art devices, become simultaneously pertinent. Such elements include [56], the manifestation of charge discreteness, the quantum electrodynamical (QED) *Casimir effect,* quantized heat flow, manifestation of the wave nature of electrons, quantum information theory, computing and communications, wave behavior in periodic and non-periodic media, and quantum squeezing. In this chapter, and the following, we expose fundamental knowledge required to analyze devices exploiting these phenomena.

2.2 Manifestation of Charge Discreteness

2.2.1 Effects of Charge Discreteness in Transmission Lines

The most fundamental element in circuits and systems is the interconnect or transmission line (TL). TLs play an essential role in configuring circuits and systems at all length scales [56]. Ideally, TLs are the medium through which signals propagate, from one point to another, with no effect on the signals, except a frequency-independent delay. Figure 2-1 shows a sketch of a microstrip TL, a commonly used TL in integrated circuits. It consists of a metallic stripe of width w and thickness t_s, patterned on a dielectric substrate of thickness h and dielectric constant ε_r, with the substrate resting on a metallic ground plane.

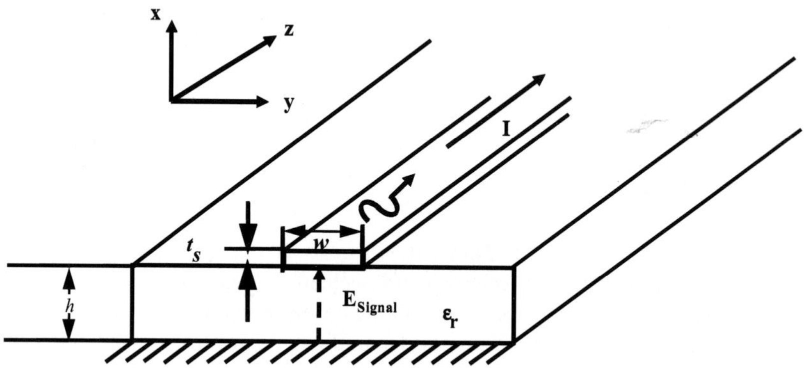

Figure 2-1 Sketch of microstrip transmission line.

From an electromagnetics perspective, the TL's qualitative operation is simple [57]. The signal of interest is impressed at its input, by way of its equivalent electric field E_{Signal} between the metallic stripe and the ground plane, and it elicits a propagating quasi-TEM electromagnetic wave which is guided in the dielectric substrate region between the stripe and the ground plane. A current *I*, flowing in one direction in the stripe, and in the opposite direction in the ground plane, embodies the boundary conditions necessary to sustain the propagating wave in the substrate, as per Maxwell's equations [57], and the magnitudes of the magnetic and electric fields stored along the line give rise to an inductance per unit length, *L*, and a capacitance per unit length, *C*, whose ratio is captured in the so-called characteristic impedance of the line, given by $Z_0 = \sqrt{L/C}$. TLs are usually designed to have $Z_0 = 50\Omega$, which results if, for example, $h = 635\mu m$, $w = 635\mu m$,

$t_s = 2\mu m$, and $\varepsilon_r = 9.8$. Under these conditions of a metal stripe of relatively large dimensions with respect to a Bohr radius, a_0=0.592Å=0.0592nm, the current I may be construed as consisting of an ensemble of freely-propagating electrons, each characterized by a plane wave-*like* wave function $\psi \sim e^{ikz}$, with *continuous* energy $E = \hbar^2 k^2 / 2m^*$, where $\hbar$ is Planck's constant, $k = 2\pi/\lambda$ is the wave vector, λ the electron wavelength, and m^* the effective mass [58].

Assuming a lossless TL, its circuit behavior may be represented as a tandem connection of a number of finite-length cells, each cell consisting of a length Δz of its inductance, L, and capacitance, C, per unit length, see Figure 2-2(a) [56].

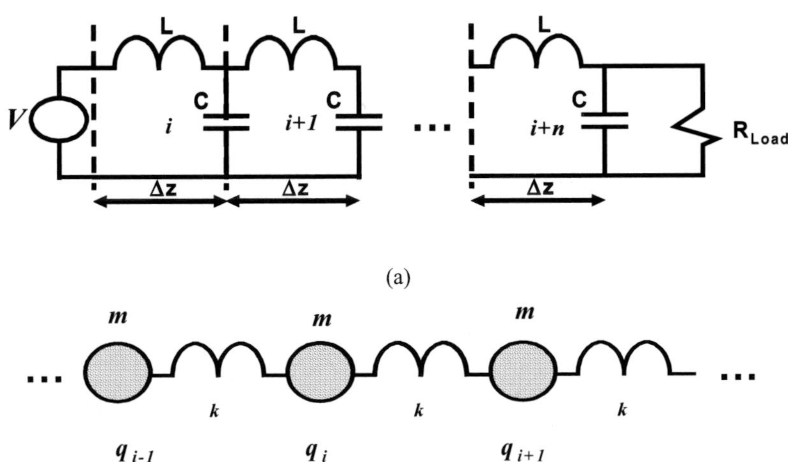

(b)

Figure 2-2. (a) Model of ideal transmission line. (b) Model of monatomic linear chain.

Thus, the propagation of a signal from a source towards a load, down a TL, can be visualized as an advancing tide of *charge fluid* charging the successive cells until the load is reached.

Enter nanotechnology. In concert with exploiting the ability to pattern nanoscale circuits, it is expected that TLs with stripes of nanoscale and sub-nanoscale widths and thicknesses will be prominent. In this context, electron currents will be transported down very narrow and thin metallic wires, so narrow and thin, in fact, that their dimensions may stop at only tens of Bohr radii. This means that the electrons involved will not only experience quantum mechanical confinement, i.e., that their energy will become quantized and given by [58], [59]:

$$E \sim \frac{\hbar^2}{2m^*}\left[\left(\frac{\pi n_x}{t_s}\right)^2 + \left(\frac{\pi n_y}{w}\right)^2 + k_z^2\right] \tag{1}$$

but also, that their discrete nature will be manifest. This latter feature becomes operative when the system size along a transport dimension becomes of the order of the carrier inelastic coherence length, and it implies that, in addition to the quantum mechanical energy of confinement of Eq. (1), the Coulomb charging energy required for adding or removing an electron, $E_c = q^2/L_i$ where L_i is a characteristic length in direction *i*, must be taken into account [58-62]. One must then turn to quantum mechanics to properly describe the TL behavior.

The observation [61]-[63], that the charge q in successive cells, and the total energy, obey equations (2) and (3),

$$L\frac{d^2 q_i}{dt^2} = \frac{1}{C}(q_{i+1} + q_{i-1} - 2q_i) \tag{2}$$

$$H = \sum_i \left(\frac{1}{2L}\left(L\frac{dq_i}{dt}\right)^2 + \frac{1}{2C}(q_{i+1} - q_i)^2\right) \tag{3}$$

whose forms are *identical* to the equations describing the longitudinal vibration modes in a monatomic linear chain (MLC) [64] (see Appendix A), Figure 2(b), motivated the application of the quantum mechanical description of the latter to the TL. In particular, in (3), the first and second terms account for the magnetic and electric energies in the TL inductors and capacitors, respectively, and $p = L\frac{dq}{dt}$ and q play the roles of "momentum" and "coordinate," respectively. Notice, however, that since q is charge, p represents electric current.

The above TL quantization assumed the electric charge q to be a *continuous* variable. As has been observed [59], however, under appropriate circumstances, e.g., system size close to the inelastic coherence length, the particle (or *discrete*) nature of electrons becomes evident. Li [61] considered the consequences of this possibility and, accordingly, advanced a theory for TL quantization assuming q to be discrete.

The possibility of having the charge adopt exclusively discrete values, was introduced [61] by imposing the condition that the eigenvalues of the charge operator $\hat{q}$ be discrete, i.e.,

$$\hat{q}|q\rangle = nq_e|q\rangle \tag{4}$$

In other words, the result of measuring the charge in the TL must be n times the fundamental electron charge, q_e, where n is a positive *integer*. Since, from a comparison with the MLC description, charge adopts the role of a "coordinate" operator in the quantized Hamiltonian, the form of the corresponding "momentum" operator $\hat{p}$, and in particular,

$$\hat{p}^2 = \left(\frac{\hbar}{i}\frac{\partial}{\partial q}\right)^2 = -\hbar^2\frac{\partial^2}{\partial q^2} \tag{5}$$

must reflect this new situation. This is accomplished by replacing the partial derivative by its finite-difference approximation in charge coordinate space [65], i.e.,

$$\frac{\partial^2\psi}{\partial q^2} = \frac{\psi(n+1) - 2\psi(n) + \psi(n-1)}{q_e^2} \tag{6}$$

where q_e is the fundamental unit discretizing the charge "axis" and ψ is the electron wavefunction in the charge representation. Assuming the line is driven by a voltage source V, Schrödinger's for the TL is given by Eq.(7) [61, 62]:

$$-\frac{\hbar^2}{2q_e^2 L}\{\psi_{n+1} - 2\psi_n + \psi_{n-1}\} + \left\{\frac{\hat{q}^2}{2C} + \hat{q}V\right\}\psi_n = \varepsilon\psi_n \tag{7}$$

or, using Eq. (4):

$$-\frac{\hbar^2}{2q_e^2 L}\{\psi_{n+1} - 2\psi_n + \psi_{n-1}\} + \left\{\frac{q_e^2 n^2}{2C} + Vq_e n\right\}\psi_n = \varepsilon\psi_n. \tag{8}$$

Imposing charge discreteness, thus, turns Schrödinger's equation for a TL into a discrete, instead of a partial, differential equation.

The implications of charge discreteness are gauged from the nature of the corresponding eigenvalues and eigenvectors for this equation. Obtaining these becomes more transparent upon developing the quantum theory of mesoscopic TLs [61, 62], which we outline below following Li [61].

With $\hat{q}$ as the charge operator, instead of the conventional spatial coordinate, the corresponding conjugate variable is taken as $\hat{p}$, which then represents the current operator, instead of the usual momentum operator. The quantum mechanics of the TL then evolves from (8) and the commutation relation:

$$[\hat{q}, \hat{p}] = i\hbar . \tag{9}$$

The fact that the eigenstates of $\hat{q}$ must be specified by an integer, *n*, allows two consecutive states to be related to one another by the application of a shift operator, in particular, $\tilde{Q} \equiv e^{iq_e\hat{p}/\hbar}$. By expanding the exponential, and using (4) and (9), this shift operator may be shown to obey the commutation relations:

$$[\hat{q}, \tilde{Q}] = -q_e\tilde{Q} \tag{10}$$

$$[\hat{q}, \tilde{Q}^+] = q_e\tilde{Q}^+ \tag{11}$$

$$\tilde{Q}^+\tilde{Q} = \tilde{Q}\tilde{Q}^+ = 1 . \tag{12}$$

The shift operator, when applied to the number eigenstates defined by, $\hat{q}|n> = nq_e|n>$, produces the following new states:

$$\tilde{Q}^+|n> = e^{i\alpha_{n+1}}|n+1> \tag{13}$$

$$\tilde{Q}|n> = e^{i\alpha_n}|n-1> \tag{14}$$

where $\alpha_n s$ are undetermined phases. Therefore, (13) and (14) lead to the interpretation of the shifter operators $\tilde{Q}^+$ and $\tilde{Q}$ as ladder operators that increase and decrease the charge of the charge operator in its diagonal representation.

The quantization apparatus is completed when the completeness and orthogonality relations, and the inner product are stipulated, in this case as given by (15)-(17), respectively,

$$\sum_n |n><n| = 1 , \tag{15}$$

$$< n|m >= \delta_{nm},\tag{16}$$

$$< \phi|\psi >= \sum_{n\in Z} < \phi|n >< n|\psi >= \sum_{n\in Z} \phi^*(n)\psi(n),\tag{17}$$

where n belongs to the set of non-negative integers Z.

These relationships permit obtaining the fundamental quantum mechanical properties of the TL, namely, the eigenfunctions of the "momentum" operator $\hat{p}$, i.e., the nature of the current, and the energy spectrum.

Assuming the usual relations [53], $\hat{p}|p >= p|p >$ and $f(\hat{p})|p >= f(p)|p >$, Li [62] expands the momentum states in terms of the number states, $|p >= \sum_{n\in Z} c_n(p)|n >$ together with the shifting operation $\tilde{Q}|p >= e^{iq_e\hat{p}/\hbar}|p >$, to obtain the relationship $c_{n+1}/c_n = \exp(iq_e p/\hbar + i\alpha_{n+1})$. This, in turn, yields the momentum expansion in terms of the number states as,

$$|p >= \sum_{n\in Z} \kappa_n e^{inq_e p/\hbar}|n >\tag{18}$$

where $\kappa_n = e^{i\sum_{j=1}^{n}\alpha_j}$ and $\kappa_{-n} = e^{-i\sum_{j=1}^{n}\alpha_{-j}}$ for n>0. Making the substitution $p \rightarrow p + \hbar(2\pi/q_e)$ in the exponential of (18) yields the same state $|p >$, from where it is determined that the momentum operator $\hat{p}$ is periodic. Further progress towards obtaining the eigenstates and dispersion is attained by noticing that, if one defines new discrete derivative operators by:

$$\nabla_{q_e}\psi(n) = \frac{\psi(n+1) - \psi(n)}{q_e},\tag{19}$$

$$\overline{\nabla}_{q_e}\psi(n) = \frac{\psi(n) - \psi(n-1)}{q_e},\tag{20}$$

then Schrödinger's equation (8), may be expressed as:

$$-\frac{\hbar^2}{2q_e^2 L}\left\{\left\{\nabla_{q_e} - \overline{\nabla}_{q_e}\right\} + \left\{\frac{\hat{q}^2}{2C} + V\hat{q}\right\}\right\}\psi = \varepsilon\psi,\tag{21}$$

from where a momentum operator $\hat{P}$, given by:

$$\hat{P} = \frac{\hbar}{2i}\left(\nabla_{q_e} + \overline{\nabla}_{q_e}\right) = \frac{\hbar}{2iq_e}\left(\widetilde{Q} - \widetilde{Q}^{+}\right), \tag{22}$$

may be defined. This new momentum operator is related to $\hat{p}$ in that $\hat{p} = \lim_{q_e \to 0} \hat{P}$.

2.2.1.1 Inductive Transmission Line Behavior

Inductive behavior is displayed by the so-called pure L-design, in which the TL is considered to have very narrow width (high impedance). Its mathematical description is given by:

$$\hat{H}_0 = -\frac{\hbar^2}{2q_e^2 L}\left\{\nabla_{q_e} - \overline{\nabla}_{q_e}\right\}, \tag{23}$$

where the terms involving the line capacitance is neglected and the driving voltage is set to zero. With this definition, and taking into account the relationship $\widetilde{Q}|p> = e^{iq_e\hat{p}/\hbar}|p>$, the following relationships are obtained:

$$\hat{P}|p> = \frac{\hbar}{q_e}\sin\left(\frac{q_e p}{\hbar}\right)|p>, \tag{24}$$

and

$$\hat{H}_0|p> = \frac{\hbar^2}{q_e^2 L}\left(1 - \cos\left(\frac{q_e p}{\hbar}\right)\right)|p>, \tag{25}$$

These are the desired momentum eigenstates and the energy spectrum. What is clear from (24) is that the current in a mesoscopic inductive line, given by $I = \hat{P}/L$, is periodic, becomes zero whenever $p = 2\pi\hbar/q_e$; $q_e \neq 0$, and that it is bounded by $\left(-\hbar/q_e L, \hbar/q_e L\right)$. Similarly, from (25) it is determined that the lowest energy state is degenerate at $p = n\hbar/q_e$.

Another peculiarity of mesoscopic TLs is the nature of their energy spectrum when formed into a ring in the presence of a magnetic flux ϕ. In this case, Schrödinger's becomes,

$$-\frac{\hbar^2}{2q_e^2 L}\left\{D_{q_e} - \overline{D}_{q_e}\right\}\psi = \varepsilon\psi , \tag{26}$$

where D_{q_e} and $\overline{D}_{q_e}$ are discrete derivatives that remain covariant in the presence of the magnetic flux ϕ and are defined by Li [61] as,

$$D_{q_e} \equiv e^{-\frac{q_e}{\hbar}\phi}\,\frac{\widetilde{Q} - e^{\frac{iq_e}{\hbar}\phi}}{q_e} \; ; \; \overline{D}_{q_e} \equiv e^{-\frac{q_e}{\hbar}\phi}\,\frac{e^{\frac{-iq_e}{\hbar}\phi} - \hat{Q}^{+}}{q_e}. \tag{27}$$

Applying the Hamiltonian in (26) to the eigenstate $|p>$, the energy eigenvalues are obtained as,

$$\varepsilon(p,\phi) = \frac{2\hbar}{q_e^2}\sin^2\left(\frac{q_e}{2\hbar}(p-\phi)\right), \tag{28}$$

where ϕ is the magnetic flux threading the TL. Thus, (28) implies that when the discrete nature of charge is at play, the TL energy becomes a periodic function of p or ϕ, with maximum amplitude $\frac{2\hbar}{q_e^2}$ and nulls occurring whenever $p = \phi + n\hbar/q_e$. Furthermore, it has also been shown that the TL current is given by,

$$I(\phi) = \frac{\hbar}{q_e L}\sin\left(\frac{q_e}{\hbar}\phi\right), \tag{29}$$

which implies that it becomes an oscillatory function of the magnetic flux. Since no applied forcing function was assumed, (29) leads to the important observation [62] that a TL in the discrete charge regime will, in the presence of a magnetic flux, exhibit *persistent currents* [59]. These are currents without dissipation, such as the atomic orbital currents that elicit orbital magnetism.

2.2.1.2 Capacitive Transmission Line Behavior

In this design the TL is capacitive (low-impedance) and the first bracketed term in (21) is neglected and the Schrödinger equation is given by,

$$-\frac{\hbar^2}{2q_e^2 L}\left\{\frac{\hat{q}^2}{2C}+V\hat{q}\right\}\psi=\varepsilon\psi, \tag{30}$$

In this case, the Hamiltonian operator commutes with the charge operator $\hat{q}$, and consequently [60], they have simultaneous eigenstates. In particular, the energy of the state $|n>$ is given by [67],

$$\varepsilon=\frac{1}{2C}(nq_e-CV)^2-\frac{C}{2}V^2, \tag{31}$$

where n is the number of elemental charges describing the TL state. Thus, (31) implies that when the discrete nature of charge is at play in a low-impedance line, the TL energy is a quadratic function of the state n of charges.

An interesting phenomena is predicted for the current flow. In particular, as the applied voltage increases, the TL charge can only increase in discrete steps which are a multiple of q_e. Since the voltage required to cause this charge to be injected into the TL is q_e/C, it can be said that the voltage axis is quantized in units of q_e/C. Thus, the total charge of a line in the ground state is given by [67],

$$q=\sum_{k=0}^{\infty}\left\{u\left[V-\left(k+\frac{1}{2}\right)\frac{q_e}{C}\right]-u\left[-V-\left(k+\frac{1}{2}\right)\frac{q_e}{C}\right]\right\}q_e, \tag{32}$$

where *u(z)* is the unit step function. Consequently, by taking the time derivative of (32), one obtains the corresponding current as,

$$I=\frac{dq}{dt}=\sum_{k=0}^{\infty}q_e\left\{\delta\left[V-\left(k+\frac{1}{2}\right)\frac{q_e}{C}\right]+\delta\left[V+\left(k+\frac{1}{2}\right)\frac{q_e}{C}\right]\right\}\frac{dV}{dt}. \tag{33}$$

Eqn. (33) indicates that the current exhibits a series of *delta-function* impulses with periodicity q_e/C, consistent with every time a single electron charge is added, and amplitude proportional to the slope of the voltage source. This leads to the important observation [67] that a low-impedance ideal TL in the discrete charge regime will exhibit current flow dominated by *Coulomb blockade.*

Clearly, as limiting cases, typifying the behavior of ideal high- and low-impedance TLs in the discrete charge regime, the phenomena of persistent currents and Coulomb blockade-type current flow, respectively, raise serious questions in the context of achieving low-noise analog and reliable digital circuits and systems at nanometric-length scales. As a result, complete awareness of the possibility that these features might be inadvertently included in the design space must be incorporated in TL/interconnect models utilized in the design and analysis of future NanoMEMS.

2.2.2 Effects of Charge Discreteness in Electrostatic Actuation

One of the distinguishing features of NanoMEMS is the inclusion of functions based on mechanical structures that can be actuated. For a variety of reasons, in particular, its compatibility with IC processes, electrostatic actuation is the actuation mechanism of choice for these devices [48], and is the one on which we focus our attention next.

2.2.2.1 Fundamental Electrostatic Actuation

Perhaps the most fundamental electrostatically-actuated elements/building blocks are the singly-(cantilever) and doubly-anchored beams [52], Figure 3.

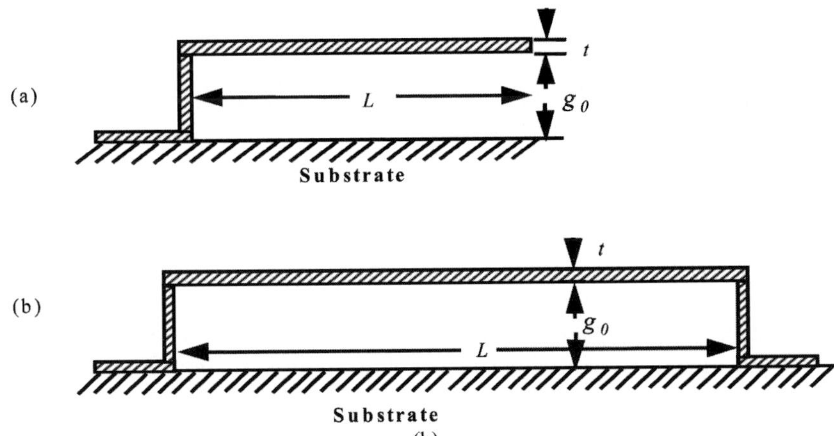

Figure 2-3. (a) Cantilever beam. (b) Doubly-anchored beam.

The devices are essentially parallel-plate capacitors, of nominal plate separation g_0, in which the top plate (beam) is free to move in response to an electrostatic force developed between it and the rigid bottom plate, as a result of a voltage applied between the two.

2.2.2.1.1 Large-signal Actuation—Switch

For typical dimensions employed in MEMS [48], e.g., beam gaps, lengths, widths, and thicknesses of about $2\mu m$, $100-250\mu m$, $10's$ of μm, and $1-10\mu m$, respectively, the displacement behavior of the beams, which manifests itself as continuous gap reduction versus applied voltage, is dictated by the equilibrium $F_{Coulomb} + F_{Spring} = 0$ established between the quadratic electrostatic force, $F_{Coulomb} = \frac{1}{2}\frac{\varepsilon_0 AV^2}{(g_0 + z)^2}$, and the linear spring force, $F_{Spring} = -k_{Beam} z$, (Hooke's law) which attempts to bring the beam back to its undeflected position. This dynamic equilibrium, and its accompanying smooth displacement, is maintained up to about one-third of the beam-to-substrate distance, at which point it is lost and the beam collapses onto the bottom plate, abruptly reducing the gap to zero. The voltage demarcating these two regimes is called *pull-in* voltage and is given by [49],

$$V_{Pull-in} = \sqrt{\frac{8k_{Beam} g_0^{\ 3}}{27A\varepsilon_0}}, \tag{34}$$

where k_{Beam} is the spring constant of the beam, and A is the electrode area.

2.2.2.1.2 Small-signal Actuation—Resonator

For application as resonators [54], an AC voltage, together with a so-called DC polarization voltage, introduced to enhance the current elicited by the variable beam capacitance, are applied. Since the resonators are intended for application as stable frequency standards, with frequency given by [18],

$$f_{r,nom} = 1.03\kappa\sqrt{\frac{E}{\rho}}\frac{h}{L_r^2}, \tag{35}$$

where κ is a scaling factor that models the effects of surface topography, including for instance, the anchor step-up and its corresponding finite elasticity, E is the Young's modulus of the beam material, ρ its density, h its thickness and L_r its length, the combined amplitude of AC and DC voltages is chosen to be lower than pull-in, thus keeping the beam from collapsing.

2.2.2.2 Coulomb Blockade

The phenomenon of Coulomb blockade [68, 69] refers to the fact that under certain conditions, namely, when junctions are defined whose capacitance is of the order of $C \sim 10^{-15} F$ or less, the energy required to increase the charge by one electron is not negligible with respect to temperature. For example [68], Figure 2.3 shows that, while a neutral metallic island, such as the plates of a capacitor, emits no electric field and, thus, allows the unimpeded approach of an electron, once this electron becomes part of the island it emits an electric field that may prevent the addition of more electrons.

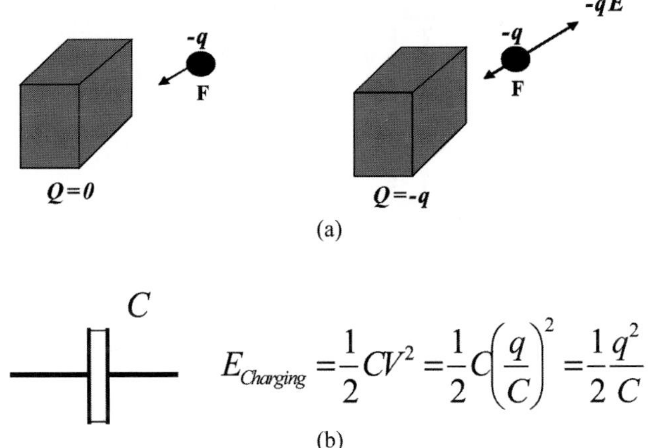

C

$$E_{Charging} = \frac{1}{2}CV^2 = \frac{1}{2}C\left(\frac{q}{C}\right)^2 = \frac{1}{2}\frac{q^2}{C}$$

(b)

Figure 2-4. (a) Charging Coulomb island. (a) Charging energy of small capacitor.

At this point, the island *blocks* such an addition of extra charge. For a junction capacitance of $C \sim 10^{-15} F$, the minimum voltage required to add a charge q is q/C, thus the charging energy is $E_C = q^2/2C = 1.283 \times 10^{-23} J$, which is close to the thermal energy at 1K. If the capacitance were smaller, e.g., $C \sim 6.2 \times 10^{-18} F$, such as might be typical for nanoparticles, then the charging energy would be close to the thermal energy at 300K. The implication of this is that it may be impossible to continuously inject charges into the capacitor when the charging energy exceeds the ambient temperature. Rather, for an increasing applied voltage, a charging event only occurs every time its magnitude exceeds the charging energy of an electron; one enters the Coulomb blockade regime and the current into the capacitor becomes pulse-like. The situation is illustrated in Figure 2-4 with respect to the so-called single-electron box [69].

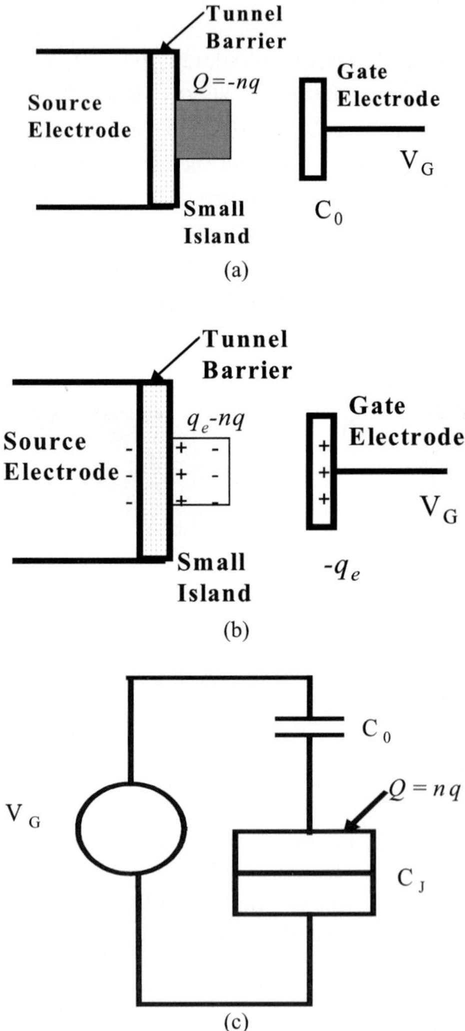

Figure 2-5. Voltage-controlled electron injection into metallic island. (a) V_G=0. (b) V_G>V_C. (c) Circuit model (*After* [68], [69].)

A voltage source V_G is connected through a small capacitor C_0, to a small metallic island that rests over a tunnel barrier which, in turn, is in contact with an electron reservoir. The capacitance of the tunnel barrier is denoted C_J, and the distance between the gate electrode and the small island, defining C_0, is such that tunneling is suppressed [69]. With V_G=0, the system is neutral; the small island containing *n* positive charges *q*, which are neutralized by an equal amount of negative charges *-nq*, Figure 2-24(a). When the gate voltage increases, the number of electrons in the small island may change by amounts $q_e = C_0 V_G$, Figure 2-4(b). In particular, the field induced by the gate causes an uncompensated charge *nq* to appear on the island. The capacitance "seen" by the island is C_0+C_J. Therefore, the charging energy accompanying the injection of a charge $q_e = C_0 V_G$ is,

$$E_C = \frac{(nq - q_e)^2}{2(C_0 + C_J)}, \tag{36}$$

It is noticed that, while the external charge q_e is continuous, the island charge may only increase in discrete steps of value *q*. Therefore, the island charge is a step-like function of the gate voltage. As a function of temperature, the average number of electrons in the island is given by [68] (37), Figure 2-5.

$$\langle n \rangle = \frac{\sum_{-\infty}^{\infty} n e^{-E_C/k_B T}}{\sum_{-\infty}^{\infty} e^{-E_C/k_B T}} \tag{37}$$

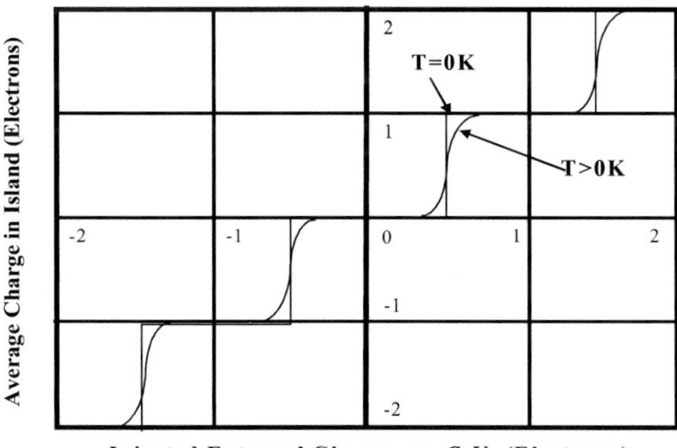

Figure 2-6. Average island charge versus injected charge. (*After* [69].)

2.2.3 Single-Electron Tunneling

Upon the island being populated by the injected charge, the charge tunnels through C_J and diffuses to the leads in a characteristic time τ given by the uncertainty principle (38) [69].

$$E_C \geq \frac{\hbar}{\tau}, \tag{38}$$

If the bias V_G causes the injection of a charge q every τ seconds, then a current of magnitude $I = q / \tau$ is set up, Figure 2-7.

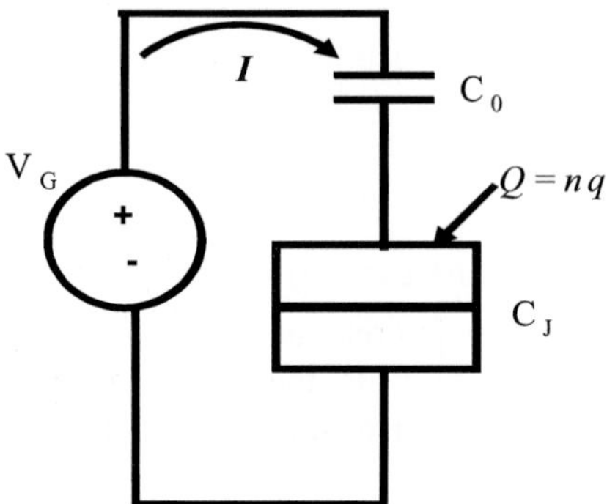

Figure 2-7. Single-electron tunneling schematic.

However, if this time is too short, then the current would appear to be continuous, as opposed to pulse-like. In this case, no discrete, single-electron tunneling event is observed. To observe single-electron tunneling, the characteristic time must exceed the product of the capacitance times the lead resistance, $\tau > RC$, a condition which leads to a minimum value for lead resistance, Eq. (38).

$$R > \frac{2\hbar}{q^2} \tag{38}$$

Notice that transport is occurring through a tunneling junction.

2.2.3.1 Quantum Dots

Quantum Dots (QDs) are structures in which electrons are confined in all three dimensions [59]. These structures include both gated layered structures grown by MBE, and metal and semiconductor nanoparticles up to several nanometers, e.g., ~1-6 nm, in size. Because of their small size, which is comparable to that of the Bohr exciton, $a_{ex} = \varepsilon \cdot \hbar^2 / m_{ex} e^2$, electron energy levels in QDs are quantized. Electron transport through a QD is mediated by tunnel barriers, see Fig. 2-8, and is effected via a series of individual tunneling events across the barriers.

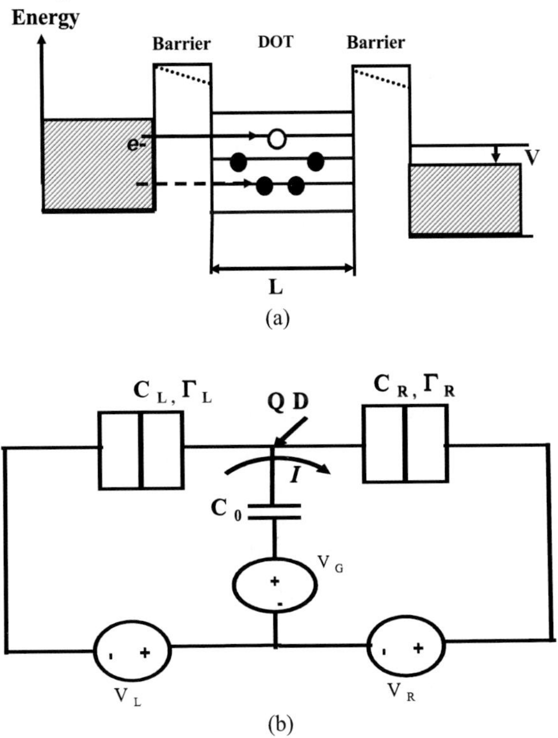

Figure 2-8. (a) Sketch of quantum dot energy level diagram. The continuous line denotes equilibrium, while the dashed line denotes reflects an applied voltage, V. The dashed arrow denotes suppressed current due to Coulomb blockade by QD electrons. (b) Equivalent circuit of QD.

The tunneling rate across the barriers is characterized by the change in free energy, Δ, resulting from the tunneling event, and the tunnel resistance, R_t ($R_t >> h/e^2$), and is given by [70], [71] Eq. (39).

$$\Gamma = \frac{\Delta}{e^2 R_t \left(1 - \exp\left(\frac{-\Delta}{k_B T}\right)\right)}. \qquad (39)$$

In general, the tunneling rate will depend on the number of available (empty) states within the QD. If Γ_f is the tunneling rate into level *f* in the QD, g_f is the degeneracy factor, m_f is the number of electrons already occupying the level, and $F(\varepsilon) = 1/(1 + \exp(\varepsilon/k_B T))$ is the Fermi function, then the total tunneling rate is given by,

$$\Gamma_{QD}^{FS} = \sum_f \Gamma_f \cdot (g_f - m_f) F(\varepsilon_f^{QD} - \Delta), \tag{40}$$

where the initial and final electron energies are related by, $\varepsilon_i^{FS} = \varepsilon_f^{QD} - \Delta$, ε_i^{FS} being the initial electron energy [232], [233]. Notice that, at small bias voltages, the occupancy of QD states precludes tunneling due to Coulomb blockade.

2.2.4 Quantized Electrostatic Actuation

In contrast to conventional electrostatically-actuated MEM devices, which exhibit continuous displacement versus bias behavior prior to pull-in, the advent of precision nanoelectromechanical fabrication technology [72] and carbon nanotube synthesis [17] has enabled access to beams with dimensional features (gaps, lengths, widths, and thicknesses) of the order of several hundred nanometers in which conditions for the manifestation of charge discreteness become also evident. In fact, recent [73] theoretical studies of suspended (doubly anchored/clamped) carbon nanotubes (CNTs) in which Coulomb blockade dominates current transport have predicted that charge quantization in the CNTs will result in quantization of their displacement.

Specifically, Sapmaz, *et al.* [73] considered a single-wall nanotube (SWNT) modeled as a rod of radius *r*, and length *L*, and separated by a gap g_0 over a bottom electrode, Fig. 2-9.

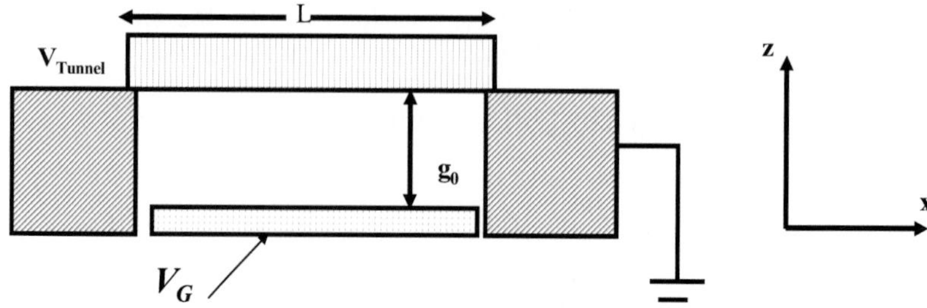

Figure 2-9. Schematic of suspended CNT as doubly anchored beam.

They described its behavior as follows. As the actuation voltage, V_G, applied between the CNT and the bottom electrode increases, the beam bends downwards causing the applied electrostatic energy to be converted into elastic deformation energy, given by,

$$U_{Elastic}[z(x)] = \int_0^L dx \left\{ \frac{EI}{2} z''^2 + \frac{\hat{T}}{2} z'^2 \right\}, \tag{41}$$

where E and $I = \pi r^4/4$ are the CNT Young's modulus and moment of inertia, respectively, and $\hat{T} = T_0 + T$ is total stress, comprised of the residual stress, T_0, and the stress induced by V_G, which is given by,

$$T = \frac{ES}{2L} \int_0^L z'^2 dx, \quad S = \pi r^2. \tag{42}$$

Since, ignoring residual stress, the beam elastic energy must correspond to the electrostatic energy that induced it, the total energy the state of deformed the beam arrives at is that at which the sum of elastic and electrostatic energies is a minimum. In the Coulomb blockade regime, however, as the bias voltage V is raised, a discrete number of charges, nq, populates the suspended CNT. Thus, the electrostatic energy must include this contribution, in addition to the actuation voltage (V_G)-induced deformation. Taking both electrostatic energy sources, into account, Sapmaz, *et al.* [73] approximated the total electrostatic energy by,

$$U_{Electrostatic}(z(x)) = \frac{(nq)^2}{2C_G(z)} - nqV_G \approx \frac{(nq)^2 \ln \frac{2R}{L}}{L} - \frac{(nq)^2}{L^2 R} \int_0^L z(x)dx \tag{43}$$

then, minimizing the total energy with respect to z, the following equation for the CNT bending was obtained,

$$IEz'''' - Tz'' = F_0 = \frac{(nq)^2}{L^2 R}. \tag{44}$$

where F_0 is the electrostatic for per unit length. The bending of the doubly-anchored CNT, with the boundary conditions $z(0) = z(L) = z'(0) = z'(L) = 0$ was given as,

$$z(x)=\frac{F_0L}{2T\xi}\left[\begin{array}{l}\frac{\sinh \xi L}{\cosh \xi L-1}(\cosh \xi L-1)- \\ \sinh \xi x+\xi x-\xi\frac{x^2}{L}\end{array}\right], \quad \xi=\sqrt{\frac{T}{EI}}. \tag{45}$$

Finally, the effects of charge discreteness are manifest upon examining the maximum displacement as a function of actuation voltage, and given by (45) and (46).

$$z_{\max}=0.013\frac{(nq)^2L^2}{Er^4g_0}, \quad T \ll \frac{EI}{L^2} \quad \left(n \ll \frac{Er^5g_0}{q^2L^2}\right); \tag{46a}$$

$$z_{\max}=0.24\frac{(nq)^{2/3}L^{2/3}}{E^{1/3}r^2g_0^{1/3}}, \quad T \gg \frac{EI}{L^2} \quad \left(n \gg \frac{Er^5g_0}{q^2L^2}\right) \tag{46b}$$

$$n=Int\left(\frac{V_GL}{2r\ln(2g_0/r)}+\frac{1}{2}+\delta n\right). \tag{47}$$

For a given applied voltage, (47) gives the value of n that minimizes the total energy, where δn is a small correction. Clearly, (45)-(47) reveal that the beam displacement is quantized, i.e., its position changes in discrete steps every time an electron tunnels into it.

2.3 Manifestation of Quantum Electrodynamical Forces

When the proximity between material objects becomes of the order of several nanometers, a regime is entered in which forces that are *quantum mechanical* in nature [74-76], namely, *van der Waals* and *Casimir forces*, become operative. These forces supplement, for instance, the electrostatic force in countering Hooke's law to determine the beam actuation behavior. They also may be responsible for stiction [77], i.e., causing close by elements to adhere together and, thus, may profoundly change actuation dynamics.

2.3.1 van der Waals Force

van der Waals forces, of electromagnetic and quantum mechanical origin, are responsible for intermolecular attraction and repulsion. When adjacent materials [78] are separated by distances $R>>r$, where r is the atomic radius, the wave functions decay exponentially and no bonding forces are operative. At these distances, each molecule (atom) may be characterized as a dipole antenna emitting a fluctuating field with a frequency distribution characterized by an average frequency ϖ. For distances, R, smaller than the average emitted wavelength, i.e., $R<\bar{\lambda}$ or $\frac{R\varpi}{c}<<1$, the emitted fields are reactive in nature, i.e., they vary with distance as $\vec{E}\propto 1/R^3$. Therefore, with reference to two emitting molecules (atoms), separated a distance R and endowed with dipole operators $<\hat{d}_\omega>=\alpha E_\omega$, the van der Waals interaction energy between them derives from the self-consistent field induction at each others' site. In particular, atom 1 induces a field at the site of atom 2 given by, $\hat{E}_1^{ind}(2)\approx \hat{d}_1/R^3$, which, in turn, induces a dipole at the site of atom 2 given by, $\hat{d}_2^{ind}=\alpha_2(\omega)\cdot \hat{d}_1/R^3$, where $\alpha_2(\omega)$ is the polarizability at the site of atom 2. Similarly, the induced dipole at atom 2 induces a field at the site of atom 1 given by, $\hat{E}_2^{ind}(1)\approx \frac{\hat{d}_2}{R^3}\approx \alpha_2(\omega)\cdot\frac{\hat{d}_1}{R^6}$. Thus, the average ground state dipole energy of atom 1 is given by [78], $U_\omega(R)=<\hat{d}_1^*\cdot \hat{E}_1^{ind}>=\frac{\alpha_2}{R^6}<\hat{d}_1\cdot\hat{d}_1^*>$ and is a function of its average dipole fluctuation. The signature of van der Waals forces is the $F_{vdW}=dU_{vdW}/dR\propto 1/R^7$ distance dependence.

For calculations, Desquesnes, Rotkin, and Aluru [79] have modeled the van der Waals energy by the expression,

$$U_{vdW}(R)=\int_{V_1}\int_{V_2}\frac{n_1 n_2 C_6}{R^6(V_1,V_2)}dV_1 dV_2, \tag{48}$$

where V_1 and V_2 embody two domains of integration of the adjacent materials, n_1 and n_2 are the densities of atoms pertaining to the domains V_1 and V_2, $R(V_1,V_2)$ is the distance between any point in V_1 and V_2, and C_6, with units $\left[eV\text{Å}^6\right]$, is a constant characterizing the interaction between atoms in materials 1 and 2. While a good first step for modeling purposes, the exclusively *pair wise* nature of the contributions embodied by (46) may not be accurate enough for tube geometry since it is known [80] that, in exact calculations, one needs to consider three-particle, four-particle, etc interactions, or equivalently multi-pole interactions. These multiple interactions must be included to improve modeling results. Nevertheless, applied to a SWNT beam of diameter r and suspended by a gap R, they obtained the van der Waals energy per unit length of the CNT as,

$$\frac{U_{vdW}}{L} = \frac{C_6 \sigma^2 \pi^2 r(r+R)(3r^2 + 2(R+r)^2)}{2((R+r)^2 - r^2)^{7/2}}, \tag{49}$$

where $\sigma \cong 38nm^{-2}$ is the atomic surface density, L is the CNT length. The corresponding van der Waals force is given by,

$$F_{vdW} = \frac{d\left(\frac{U_{vdW}}{L}\right)}{dR} = \frac{-\left(C_6 \sigma^2 \pi^2 r \sqrt{R(R+2r)}\right) \cdot \left(8R^4 + 32R^3 r + 72R^2 r^2 + 80Rr^3 + 35r^4\right)}{2R^5 (R+2r)^5}. \tag{50}$$

As mentioned previously, the van der Waals force is one contributor to the phenomenon of stiction. Thus, its prominence must be accounted for in the design of advanced structures, e.g., nanoelectromechanical *frequency tuning* systems [54] based on *quantum gears* [81], as estimates of its magnitude are useful in designing against it [18, 82].

2.3.2 Casimir Force

The Casimir force arises from the polarization of adjacent material bodies, separated by distances of less than a few microns, as a result of *quantum-mechanical fluctuations* in the electromagnetic field permeating the free space between them [74-77]. It may also arise if vacuum fluctuations are a classical real electromagnetic field [83]. The force may be computed as retarded van der Waals forces or as due to changes in the boundary conditions of vacuum fluctuations; these are equivalent viewpoints as far as it is known [80].

When the material bodies are parallel conducting plates, separated by free space, the Casimir force is attractive [74], however, in general whether the force is attractive or repulsive [82], [84] depends on both the boundary conditions, including specific geometrical features, imposed on the field as well as the relationship among material properties of the plates and the intervening space. For example, repulsive forces are predicted by

Lifshitz formula [75] if the material between two plates has properties that are intermediate between those of the plates.

The startling aspect of the Casimir force is that it is a manifestation of the purely quantum-mechanical prediction of *zero-point vacuum fluctuations* [74-77] (see Appendix A), i.e., of the fact that, even in circumstances in which the average electromagnetic field is zero, its average energy shows fluctuations with *small* but non-zero value, i.e., *there is virtually infinite energy in vacuum.* Research efforts aimed at the practical exploitation of this extremely large energy source, residing in free space, are under way [85-87].

Calculating the Casimir force entails circumventing the fact that the zero-point vacuum energy, $E_{Field} = \frac{1}{2}\hbar\sum_n \omega_n$ diverges, and many techniques to accomplish this have been developed [74-77], [88], [89], but including these in our presentation is well beyond the scope of this article. The essence of many of these calculations, however, is to compute the physical energy as a difference in energy corresponding to two different geometries, e.g., the parallel plates at a distance "*a*" apart, and these at a distance "*b*," where the limit as *b* tends to infinity is taken. For flat surfaces, the infinite part of the energy cancels when the energy difference of the two configurations is taken. The calculated zero-temperature Casimir energy for the space between two uncharged perfectly conducting parallel plates, Figure 2-10,

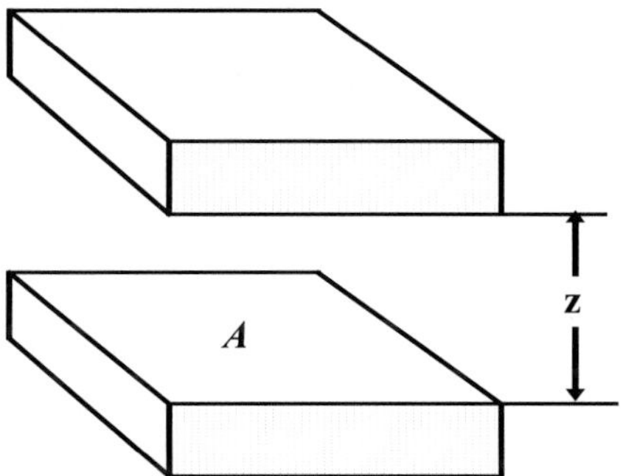

Figure 2-10. Casimir effect geometry.

is given by,

$$U_{Casimir}(z) = -\frac{\pi^2 \hbar c}{720}\frac{1}{z^3}, \tag{51}$$

and, the corresponding Casimir force per unit area is given by,

$$\frac{F_{Cas}^{0}}{A}=-\frac{\pi^{2}\hbar c}{240}\frac{1}{z^{4}}. \tag{52}$$

For planar parallel metallic plates with an area $A=1cm^2$ and separated a distance $z=0.5\mu m$, the Casimir force is $2\times10^{-6}N$.

Many experiments measuring the Casimir force under various conditions, such as effecting normal displacement between a sphere and a smooth planar metal and between parallel metallic surfaces, as well as, effecting lateral displacement between a sphere and a sinusoidally corrugated surface, have been performed [89-95]. A good recent review of experiments and theory for Casimir forces has been published by Bordag, Mohideen, and Mostepanenko [89].

Since the Casimir energy/force is a sensitive function of the boundary conditions, corrections to the ideal expression (52) have been introduced to account for certain deviations. For example, for the sphere-plate geometry, the zero-temperature Casimir force is given by,

$$F^{0}_{Cas\_Sphere-Plate}\left(z\right)=-\frac{\pi^{3}}{360}R\frac{\hbar c}{z^{3}}, \tag{53}$$

where R is the radius of curvature of the spherical surface.

To include the finite conductivity of the metallic boundaries, two approaches have been advanced. In one, the force is modified as [96, 97],

$$F_{Cas}^{0,\sigma}\left(z\right)=F^{0}_{Cas\_Sphere-Plate}\left(z\right)\left[1-4\frac{c}{z\omega_{p}}+\frac{72}{5}\left(\frac{c}{z\omega_{p}}\right)^{2}\right], \tag{54}$$

where ω_p is the metal plasma frequency [64]. In the other, obtained by Lifshitz [98], the correction is ingrained in the derivation of the Casimir force, and is given by,

$$F_{Cas}^{0,\sigma}\left(z\right)=-\frac{R\hbar}{\pi c^{3}}\int_{0}^{z}dz'\int_{0}^{\infty}\int_{1}^{\infty}p^{2}\xi^{3}dpd\xi\times\left\{\begin{array}{l}\left[\frac{\left(s+p\right)^{2}}{\left(s-p\right)^{2}}e^{\frac{2p\xi z}{c}}-1\right]^{-1}+\\\left[\frac{\left(s+p\varepsilon\right)^{2}}{\left(s-p\varepsilon\right)^{2}}e^{\frac{2p\xi z}{c}}-1\right]^{-1}\end{array}\right\}, \tag{55}$$

where $s=\sqrt{\varepsilon-1+p^2}$, $\varepsilon(i\xi)=1+\frac{2}{\pi}\int_0^\infty \frac{\omega\varepsilon''(i\xi)}{\omega^2+\xi^2}d\omega$ is the dielectric constant of the metal, ε'' is the imaginary component of ε, and ξ is the imaginary frequency given by $\omega=i\xi$.

Corrections due to nonzero temperature yield [77],

$$F_{Cas}^{T}(z)=F_{Cas}^{0}(z)\left[1+\frac{720}{\pi^2}f(\zeta)\right], \tag{56}$$

where $\zeta=k_BTz/\hbar c$, k_B is Boltzmann constant, T is the absolute temperature, and

$$f(\zeta)\approx\begin{cases}\left(\zeta^3/2\pi\right)\vartheta(3)-\left(\zeta^4\pi^2/45\right), & for \quad \zeta\le 1/2\\ \left(\zeta/8\pi\right)\vartheta(3)-\left(\pi^2/720\right), & for \quad \zeta>1/2\end{cases}, \tag{57}$$

with $\vartheta(3)=1.202...$.

Roy and Mohideen [90] included originally the effects of surface roughness, which changes the surface separation, by replacing the flat plate with a spatial sinusoidal modulation of period λ, and the energy averaged over the size of the plates, L, to obtain,

$$<U_{Casimir}\left(z+A\sin\frac{2\pi x}{\lambda}\right)>=-\frac{\pi^2\hbar c}{720}\frac{1}{z^3}\sum_m C_m\left(\frac{A}{z}\right)^m, \tag{58}$$

where A is the corrugation amplitude. The corresponding Casimir force is then given by the so-called, *Force Proximity Theorem* [99] relating the parallel plate geometry and the sphere-plate geometry, namely,

$$F_{Cas\_Roughness}=2\pi R<U_{Cas\_Rouchness}> \tag{59}$$

For $\lambda << L$ and $z+z_0>A$, where z_0 is the average surface separation after contact due to stochastic roughness of the metal coating, they recommend the following coefficients in (58): $C_0=1$, $C_2=3$, $C_4=45/8$, $C_6=35/4$. A more accurate and general model for stochastic surface roughness, advanced by Harris, Chen, and Mohideen [88], includes the effects of surface roughness, by replacing the flat plate with the mean stochastic roughness amplitude A, to obtain,

$$F_{Cas}^{r}(z) = F_{Cas}^{0}(z)\left[1 + 6\left(\frac{A}{z}\right)^{2}\right], \tag{60}$$

where A is derived from direct measurements via an Atomic Force Microscope (AFM).

2.4 Quantum Information Theory, Computing and Communications

The advent of nanoscale fabrication techniques has brought within our reach the possibility of producing systems whose predominant behavior is described by quantum mechanics (QM). While the engineering of systems based on exploting this new physics/technological paradigm is still in its infancy, this new paradigm is ultimately expected to manifest itself in the ushering of a 'new electronics' technology era. Obviously, this 'new electronics' is expected to change the way in which systems are implemented to effect the functions of information processing, computing and communications [100-111]. These functions, in turn, will exploit the properties of quantum mechanical wave functions. In this section we introduce key aspects of the fundamental physics on which these functions are predicated, in particular, we focus on the concepts underpinning quantum information processing, namely, quantum bits (*qubits*), quantum entanglement, the Einstein-Podolsky-Rosen (EPR) State, quantum gates, and quantum teleportation.

Quantum information is represented by quantum bits or *qubits* [103]. Qubits are fundamental physical entities, such as a two-level atom, which may adopt two possible quantum (stationary) states (see Appendix A), say the mutually orthogonal states $|0\rangle$ and $|1\rangle$. Due to its quantum nature, however, the most general state is expressed as,

$$|\psi\rangle = a|0\rangle + b|1\rangle, \tag{61}$$

i.e., as a superposition of both states. Thus, a measurement of the qubit will cause its wavefunction to *collapse* into the state $|0\rangle$ with probability $|a|^2$, or into the state $|1\rangle$ with probability $|b|^2$. This means that during its time evolution a qubit may be partly in both the $|0\rangle$ and $|1\rangle$ state at the same time, i.e., to the degree that *a* and *b* may adopt an infinity of values, the qubit has the potential to be in *any* of these. A quantum system possessing *n* qubits is said to have 2^n accessible mutually orthogonal quantum states. For example, a system containing two noninteracting qubits will have the four states: $|00\rangle$, $|01\rangle$, $|10\rangle$, $|11\rangle$. States such as these, which represent the juxtaposition of independent or noninteracting systems (qubits), are called *tensor product* states.

2.4.1 Quantum Entanglement

In general, a tensor product provides the mathematical description of the state of a system that is constituted by bringing together noninteracting quantum systems, assuming that they remain without interacting [60]. Comprehending this concept is useful to get a clear understanding of the definition of an *entangled* state [107-111].

In particular, if associated with two quantum systems there are vector spaces V_1 of dimension N_1, in which resides a vector $|\phi\rangle$, and V_2 of dimension N_2, in which resides a vector $|\chi\rangle$, and where N_1 and N_2 may be finite or infinite, then the tensor product of V_1 and V_2 is denoted by the vector space V [60],

$$V = V_1 \otimes V_2, \tag{62}$$

of dimension $N_1 N_2$, where the vector,

$$|\phi\rangle \otimes |\chi\rangle = |\phi\rangle |\chi\rangle, \tag{63}$$

associated with the overall space V, is called the tensor product of $|\phi\rangle$ and $|\chi\rangle$.

If the vectors $|\phi\rangle$ and $|\chi\rangle$ can be expressed in terms of the respective bases $\{|u_i\rangle\}$ and $\{|v_i\rangle\}$, so that,

$$|\phi\rangle = \sum_i a_i |u_i\rangle, \tag{64}$$

and

$$|\chi\rangle = \sum_j b_j |v_j\rangle, \tag{65}$$

then, the tensor product may be written as,

$$|\phi\rangle \otimes |\chi\rangle = \sum_{i,j} a_i b_j |u_i\rangle \otimes |v_j\rangle, \tag{66}$$

from where it is seen that the components of a tensor product vector are the products of the components of the two vectors of the product. An example will help appreciate the meaning of a tensor product immediately. Let V_x and V_y be two vector spaces in which the bases $\{|x\rangle\}$ and $\{|y\rangle\}$, reside. Then the tensor product of the spaces is given by,

$$V_{xy} = V_x \otimes V_y, \tag{67}$$

and the tensor product of the bases is given by,

$$|xy\rangle = |x\rangle|y\rangle. \tag{68}$$

Consequently, if X and Y are operators in V_{xy}, then we have,

$$X|xy\rangle = X|x\rangle(|y\rangle) = x|x\rangle(|y\rangle) = x|x\rangle|y\rangle = x|xy\rangle, \tag{69}$$

$$Y|xy\rangle = (|x\rangle)Y|y\rangle = (|x\rangle)y|y\rangle = y|x\rangle|y\rangle = y|xy\rangle. \tag{70}$$

Essentially, then, the operators acting over a tensor product of spaces operate *only* on the vector space to which they belong.

Now, assume that the global state of the system is embodied by the wavefunction $|\psi\rangle \in V = V_1 \otimes V_2$. Then, according to the above, $|\psi\rangle = |\psi_1\rangle \otimes |\psi_2\rangle$, where $|\psi_1\rangle \in V_1$ and $|\psi_2\rangle \in V_2$. A quantum system is said to be *entangled* if it is impossible to express its global state as the tensor product, i.e., $|\psi\rangle \neq |\psi_1\rangle \otimes |\psi_2\rangle$. Thus, in an entangled system, it is not possible to act on one of its vector states independently without perturbing the others. It is said then, that the states in an entangled system are *correlated.*

2.4.1.1 Einstein-Podolsky-Rosen (EPR) State

In a system with two noninteracting qubits, the global state may be expressed as [108],

$$|\psi\rangle = c_1|00\rangle + c_2|01\rangle + c_3|10\rangle + c_4|11\rangle, \tag{71}$$

where $\sum_i |c_i|^2 = 1$ and each term is the tensor product of the components of the corresponding qubits. When $c_1 = c_4 = 0$, and $c_2 = c_3 = 1/\sqrt{2}$, the resulting state,

$$|\psi_{EPR}\rangle = \frac{(|01\rangle + |10\rangle)}{\sqrt{2}}, \tag{72}$$

is called an *EPR state* [108]. The EPR state is not a tensor product of the vector states, therefore, it represents an *entangled* state; it does not belong to any of the individual vector spaces, but is a combination of them. Associated with an EPR state is the so-called Bell-state basis [108], which embodies the possible states that can result upon measuring two-state quantum systems. In particular, if $|0\rangle_1$, $|1\rangle_1$ represent the two states of particle 1, and $|0\rangle_2$, $|1\rangle_2$ represent two states of particle 2, then the measurement of their EPR pair state may result in one of four state vectors, namely,

$$|\Psi^{\pm}\rangle = \frac{(|0\rangle_1|0\rangle_2 \pm |1\rangle_1|1\rangle_2)}{\sqrt{2}}, \tag{73}$$

and

$$|\Phi^{\pm}\rangle = \frac{(|0\rangle_1|1\rangle_2 \pm |1\rangle_1|0\rangle_2)}{\sqrt{2}}. \tag{74}$$

One of the most transparent demonstrations of entanglement and its implications was the experiment by Kwiat *et al.* [107], see Figure 2-11 below. This experiment exploited the principle of type-II parametric down conversion to produce directed beams of polarization entangled photons. In type-II parametric down conversion [107] an incident laser beam pump passes through a crystal, such as beta barium borate, and can spontaneously

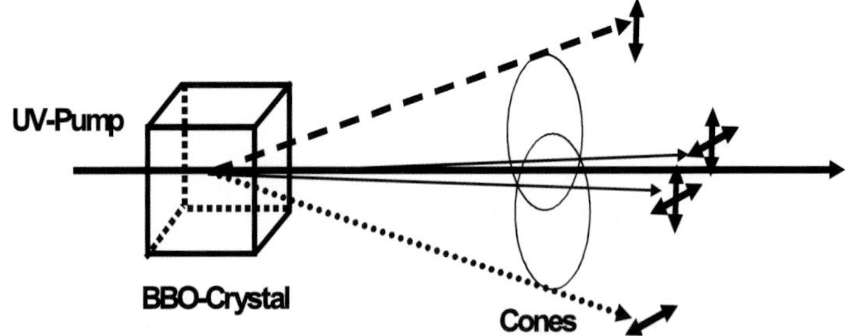

Figure 2-11. Entangled photons via type-II parametric down conversion. (*After* [107].)

decay into two photons of lower energy, one polarized vertically and one polarized horizontally, for instance. In particular, each photon can be emitted along a cone in such a way that two photons of a pair are found opposite to each other on the respective cones. If it occurs that the photons travel along the cone intersections, however, then neither photon is in a definite polarization state, but their relative polarizations are complementary, i.e., they are entangled. Taking the state of the photons along the intersecting cones as entangled, i.e.,

$$\left|\Phi^{-}\right\rangle = \frac{\left(\left|H\right\rangle_1 \left|V\right\rangle_2 - \left|V\right\rangle_1 \left|H\right\rangle_2\right)}{\sqrt{2}}, \tag{75}$$

we see that, because the polarization relationship of *complementarity* must be maintained, whenever photon 1 is measured and found to have vertical polarization, the polarization of photon 2 will be horizontal, and vice versa. This means that no matter the state in which photon 1 is found, the state of photon 2 can be predicted to be in the orthogonal state when measured. Entanglement, therefore, enables a strong correlation among the photons. This is a general property among entangled particles. By appropriately controlling the evolution of aggregates of particles, it is possible to induced them into entangled states. The agents that control the evolution of states are called *quantum gates*.

2.4.1.2 Quantum Gates

Given a qubit prepared in the initial state $\left|\psi(t_0)\right\rangle$, its state at a subsequent time t is given by $\left|\psi(t)\right\rangle = U(t,t_0)\left|\psi(t_0)\right\rangle$, where U is the qubit's *transition matrix*[60] Unitary *reversible* matrices U prescribing the evolution of qubits are called *quantum logic gates* [102], [111]. Thus, a quantum gate transforming a qubit state such that $|0\rangle \rightarrow |0\rangle$ and $|1\rangle \rightarrow \exp(i\omega t)|1\rangle$, would have the form [102],

$$U(\theta) = \begin{bmatrix} 1 & 0 \\ 0 & e^{i\theta} \end{bmatrix}, \tag{76}$$

where $\theta = \omega t$. Since U is a unitary *reversible* transformation, the quantum gate must be reversible. This means that, given the output, one must be able to uniquely determine the value of the input. There are a number of important quantum gates of which *quantum information processing* systems are made of, namely, the identity gate [100-111],

$$|0\rangle \rightarrow |0\rangle, \tag{77}$$

$$|1\rangle \rightarrow |1\rangle, \tag{78}$$

the NOT gate,

$$|0\rangle \rightarrow |1\rangle, \tag{79}$$

$$|1\rangle \rightarrow |0\rangle, \tag{80}$$

the Z gate,

$$|0\rangle \rightarrow |0\rangle, \tag{81}$$

$$|1\rangle \rightarrow -|1\rangle, \tag{82}$$

and the Hadamard gate,

$$|0\rangle \rightarrow |0\rangle + |1\rangle, \tag{83}$$

$$|1\rangle \rightarrow |0\rangle - |1\rangle. \tag{84}$$

Quantum gates are represented graphically, as in Figure 2-12 [111]. In this figure the operation of the gate is read from left to right using the following convention. Each line represents the propagation or evolution of the input state and could, accordingly, represent propagation via a wire, in time, in space, or in any other fashion evolution may be intended to take place. The gate has *control* qubits and *target* qubits. A control qubit, such as $|x\rangle$, has its line of propagation (wire) tapped at a dot. A target qubit, such as $|y\rangle$, has its line of propagation (wire) XOR'ed with a control bit. The gate's purpose is to effect a transformation on the target qubit based on the values of the control qubit, in particular, if the control qubit is set to one, then the target qubit is inverted. The realization of classical logic gates, which are inherently irreversible, by totally reversible quantum gates may be effected with the use of the Toffoli gate, see Figure 2-12(b). The Toffoli gate is an irreversible gate that takes three inputs, namely, two control qubits and one target qubit. By applying the Toffoli gate twice to its three input qubits, they are repoduced, thus the irreversible gate is made reversible [111].

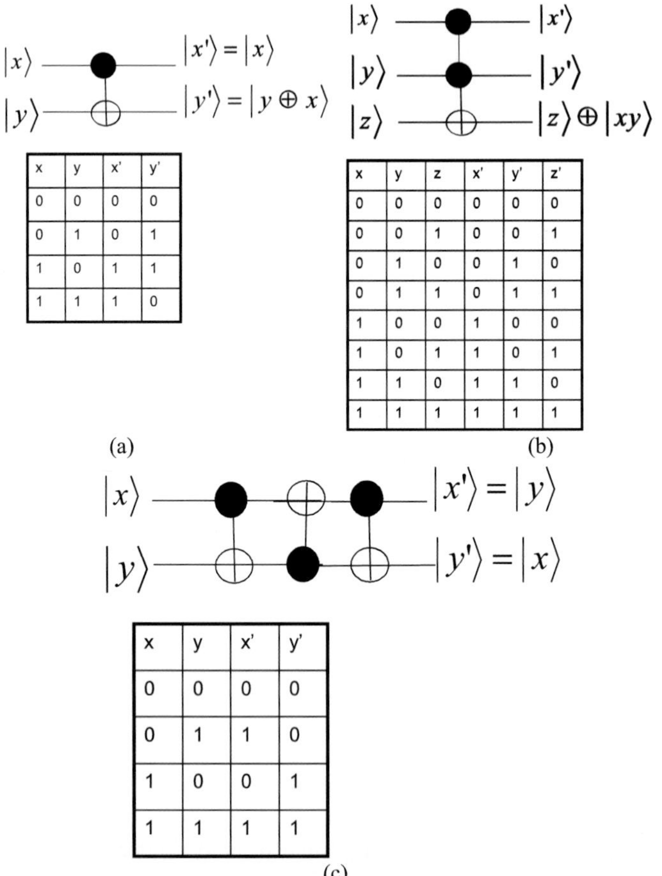

x	y	x'	y'
0	0	0	0
0	1	0	1
1	0	1	1
1	1	1	0

x	y	z	x'	y'	z'
0	0	0	0	0	0
0	0	1	0	0	1
0	1	0	0	1	0
0	1	1	0	1	1
1	0	0	1	0	0
1	0	1	1	0	1
1	1	0	1	1	0
1	1	1	1	1	1

x	y	x'	y'
0	0	0	0
0	1	1	0
1	0	0	1
1	1	1	1

Figure 2-12. Truth tables and graphical representations of some quantum gates. (a) Control-NOT gate. (b) Control-control-NOT (Toffoli) gate. (c) Bit swapping.

The control-NOT (CNOT) gate, as can be seen from Figure 2-12(a), implements the exclusive-OR (XOR) operation. Thus, the gate inverts $|y\rangle$, if $|x\rangle = 1$, and leaves it as is if $|x\rangle = 0$. This operation is expressed as,

$$C_{12}|x\rangle|y\rangle = |x\rangle|(x+y)\mathrm{mod}\,2\rangle. \tag{85}$$

Applied to a pair of single product states of two qubits, the CNOT gate produces a set of entangled qubits, i.e.,

$$C_{12}(|0\rangle_1 + |1\rangle_1)|0\rangle_2 = (|0\rangle_1|0\rangle_2 + |1\rangle_1|1\rangle_2). \tag{86}$$

Similarly, since the CNOT gate is reversible, when applied to an entangled state, it produces a set of disentangled states, i.e.,

$$C_{12}(|0\rangle_1|0\rangle_2 \pm |1\rangle_1|1\rangle_2) = (|0\rangle_1 \pm |1\rangle_1)|0\rangle_2, \tag{87}$$

and

$$C_{12}(|0\rangle_1|1\rangle_2 \pm |1\rangle_1|0\rangle_2) = (|0\rangle_1 \pm |1\rangle_1)|1\rangle_2. \tag{88}$$

These operations are essential for quantum teleportation.

One may recall that a classical NOT gate is called *universal* in the sense that any other logic gate may be created by combining several NOT gates. Similarly, a universal quantum gate should generate all unitary transformations of *n* qubits. It can be shown that such a gate is realized by combining a pair of gates, namely, one that produces a general rotation on a single bit, $U_{Universal}(\theta,\phi)$, where,

$$U_{Universal}(\theta,\phi) = \begin{bmatrix} \cos(\theta/2) & -ie^{-i\phi}\sin(\theta/2) \\ -ie^{i\phi}\sin(\theta/2) & \cos(\theta/2) \end{bmatrix}, \tag{89}$$

and a CNOT gate [100].

2.4.2 Quantum Teleportation

According to Bennett *et al.* [106], quantum teleportation is "a process that disembodies the exact quantum state of a particle into classical data and EPR correlations, and then uses these ingredients to reincarnate the state in another particle which has never been anywhere near the first particle." The process does not involve sending any qubits, rather, the sender and the receiver must have access to two other resources, namely, the ability to send classical information, and an entangled EPR pairs of particles previously shared between them.

As per the sketch of Figure 2-13, teleportation proceeds as follows.

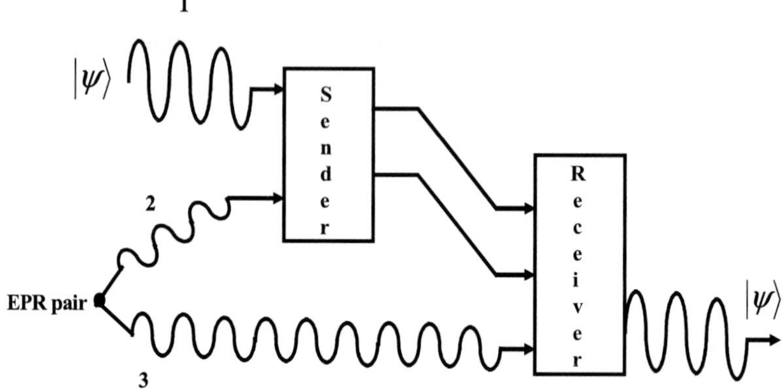

Figure 2-13. Quantum teleportation of state $|\psi\rangle$. (*After* [108].)

There are three particles involved, namely, particle 1, whose unknown state $|\psi\rangle = a|0\rangle_1 + b|1\rangle_1$ (a and b are the unknowns) is to be teleported by a *sender* to a *receiver*, and particles 2 and 3, which are prepared by an EPR source into an entangled EPR state, for instance,

$$|\Phi_{23}^+\rangle = \frac{\left(|0\rangle_2|0\rangle_3 + |1\rangle_2|1\rangle_3\right)}{\sqrt{2}}. \tag{90}$$

Of these two entangled particles, one, namely, particle 3, is sent by the EPR source to the receiver and the other, particle 2, is supplied to the sender. Notice that *locally* both the sender and the receiver possess total knowledge of the states of particles 2 and 3, respectively. However, *globally*, the three states are described by tensor product state,

$$|\psi_{123}\rangle = \frac{\left(a|0\rangle_1 + b|1\rangle_1\right)\left(|0\rangle_2|0\rangle_3 + |1\rangle_2|1\rangle_3\right)}{\sqrt{2}}, \tag{91}$$

consisting of the entangled pair, particles 2 and 3, and the unknown state. Now, the specific actions that effect the teleportation are as follows. The sender performs a joint (XOR) measurement between particles 1 and 2. As we saw previously, the outcome of measuring a pair of single product states of two qubits, such as that of particles 1 and 2, has four possible outcomes

$$\left|\Psi^{\pm}\right\rangle_{12} = \frac{\left(|0\rangle_1|0\rangle_2 \pm |1\rangle_1|1\rangle_2\right)}{\sqrt{2}}, \tag{92}$$

and

$$\left|\Phi^{\pm}\right\rangle_{12} = \frac{\left(|0\rangle_1|1\rangle_2 \pm |1\rangle_1|0\rangle_2\right)}{\sqrt{2}}. \tag{93}$$

Taking this into account, the direct product state $|\psi_{123}\rangle$ may be expanded in terms of these four outcomes and rewritten as,

$$\begin{aligned} |\psi_{123}\rangle &= \frac{\left(a|000\rangle_{123} + a|011\rangle_{123} + b|100\rangle_{123} + b|111\rangle_{123}\right)}{\sqrt{2}} \\ &= \frac{1}{2}\left|\Phi^{+}\right\rangle_{12}\left(a|0\rangle_3 + b|1\rangle_3\right) + \frac{1}{2}\left|\Phi^{-}\right\rangle_{12} \\ &\quad \cdot\left(-a|0\rangle_3 + b|1\rangle_3\right) + \frac{1}{2}\left|\Psi^{+}\right\rangle_{12}\left(b|0\rangle_3 + a|1\rangle_3\right) \\ &\quad + \frac{1}{2}\left|\Psi^{-}\right\rangle_{12}\left(-b|0\rangle_3 + a|1\rangle_3\right) \end{aligned} \tag{94}$$

The result of performing the XOR between particles 1 and 2 will be the *collapse* or projection of the *global* tensor product state $|\psi_{123}\rangle$ along one of the four vector states $\left|\Psi^{\pm}\right\rangle_{12}$ and $\left|\Phi^{\pm}\right\rangle_{12}$ with equal probability, namely, ¼. Notice that this will leave a new global state consisting of the tensor product of one of the vectors $\left|\Psi^{\pm}\right\rangle_{12}$ and $\left|\Phi^{\pm}\right\rangle_{12}$, *at the sender*, and a modified qubit 3, *at the receiver.* One possible result might be,

$$\left|\Psi^{+}\right\rangle_{12}\left(b|0\rangle_3 + a|1\rangle_3\right). \tag{95}$$

If these were the case then, to complete the teleportation process the sender has to communicate to the receiver, using classical message, that the global wave function collapsed along $\left|\Psi^{+}\right\rangle_{12}$, and that for its qubit to embody the unknown state and, hence, complete the teleportation, it has to effect the unitary transformations: $\left|0\right\rangle_{3} \rightarrow \left|1\right\rangle_{3}$ and $\left|1\right\rangle_{3} \rightarrow \left|0\right\rangle_{3}$ on its qubit 3.

2.4.3 Decoherence

A quantum system is said to decohere when, in the course of its time evolution, it loses energy to the environment. Under these circumstances its transition matrix, *U*, no longer conserves the norm of the states it acts upon. Since the states change in a random manner, the property of superposition of states is no longer maintained. From thermodynamics we know that systems that experience energy loss are irreversible, therefore, decoherence precludes the realization of quantum gates, e.g., the Toffoli gate, which must be reversible. The ability of a quantum system to maintain its coherence and, thus, be capable of manifesting superposition and entanglement, is captured by the *decoherence time*. Obviously, the system is useful for quantum information processing only during this period of time. A system made up of many qubits will exhibit a compounded amount of errors as it approaches its decoherence time., i.e., as it becomes irreversible. The decoherence of a qubit, in particular, is quantitatively captured by the *quality factor of quantum coherence* [112],

$$Q_{\varphi} = \pi \nu_{01} T_{\varphi}, \tag{96}$$

where ν_{01} is its transition frequency and T_{φ} is the coherence time of a superposition of states. While error-correcting codes techniques have been proposed to combat errors stemming from decoherence, the need for an intrinsically coherent system to begin with, remains. Therefore, the conception of approaches exhibiting long decoherence times, with respect to the intended computational function to be implemented, is crucial, if quantum information processing is to become practical. Vion *et al.* [112] point out that, given a quantum computation with elementary operations taking time t_{op}, active compensation of deciherence requires Q_{φ}'s greater than $10^{4} \nu_{01} t_{op}$. A number of approaches to the physical implementation of qubits, and their respective decoherencetimes, are discussed in Chapter 4.

2.5 Summary

This chapter has dealt with physical phenomena exploiting wave-particle duality. We began by addressing conditions that manifest charge discreteness, and its consequences on the performance of transmission lines, namely, persistent currents and current exhibiting Coulomb blockade (pulsating) behavior. Then, after introducing the concepts of single-electron tunneling, the effect of charge discreteness in electrostatic actuation was presented. In this context, we saw that charging dominated by Coulomb blockade may lead to quantized electrostatic actuation. Following this, we addressed the manifestation of quantum electrodynamical forces, in particular, van der Waals and Casimir forces and their substantial influence in moving nano- and micro-meter-scale devices. The chapter concluded with an exposition of the salient points of quantum information theory, computing and communications. In particular, we focused on the concepts of quantum bits, quantum entanglement, the Einstein-Podolsky-Rosen (EPR) state, quantum gates, and quantum teleportation. Lastly, the crucial issue of decoherence was discussed.

Chapter 3

NANOMEMS PHYSICS: QUANTUM WAVE PHENOMENA

3.1 Manifestation of Wave Nature of Electrons

The principles of nanoscale devices are based on the physics dominating this dimensional regime. In particular, as the device size is reduced below about 100nm, the electron behavior stops obeying classical physics, in which its momentum and energy are continuous, and starts obeying quantum mechanics, in which it behaves as waves with quantized energy, Figure 3-1.

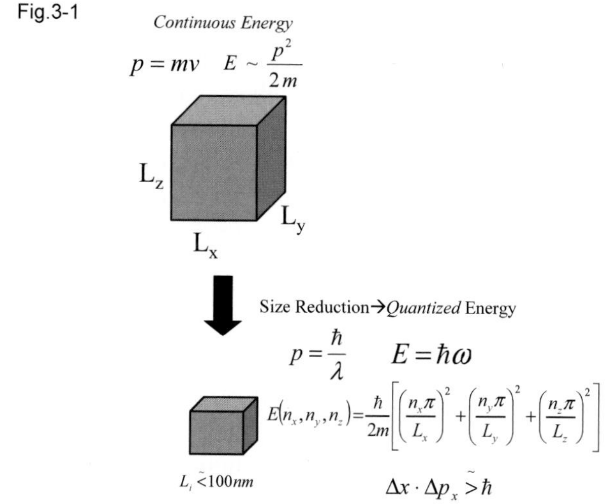

Figure 3-1. Size-dependent behavior of electrons.

Then, depending on the particular device structure, behavior such as interference, diffraction, etc., characteristic of waves, or Coulomb interaction, characteristic of particles, may be prominent. The various types of behavior are presented next.

3.1.1 Quantization of Electrical Conductance

The concept of electrical conductance quantization emerges from considering electron transport in short, narrow (quantum) wires, Figure 3-2.

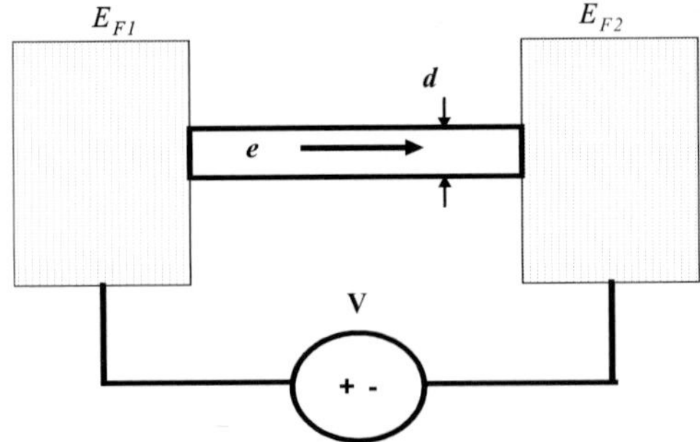

Figure 3-2. Electron transport down short, narrow wire between electron reservoirs with Fermi levels E_{F1} and E_{F2}, under the influence of applied voltage V.

Here we have a short, narrow wire connected between two electron reservoirs characterized by Fermi seas (contacts) filled up to energy levels E_{F1} and E_{F2}, Under the influence of an applied voltage V, which misaligns the Fermi levels, electrons travel from reservoir E_{F1} towards reservoir E_{F2}, in an effort to equalize the Fermi levels and, as a result, establish a current. Since the wire is very short, transport evolves without scattering, i.e., ballistically. However, since the wire is very narrow, the uncertainty principle forces its transverse momentum (and consequently, its energy) to be quantized, i.e., $p_{\perp} \sim n\hbar/d$, where *n* is an integer representing the band in which transport is occurring.

3.1.1.1 Landauer Formula

The question before us is: What is the conductance of this system? The answer was determined by Landauer [113], and may be arrived at as follows [76]. The current is the balance between the number of electrons being launched from the left-hand reservoir into the wire, and the number of electrons being launched from the right-hand reservoir into the wire. In particular, since in the momentum interval *dp*, this number equals $dp/2\pi\hbar$, the corresponding current is $dJ_+ = evdp/2\pi\hbar$. Therefore, the total left-right current, assuming a single band, and taking into account two spins, is given by:

$$J_+ = 2 \cdot \frac{e}{2\pi\hbar} \int_0^\infty v dp = \frac{e}{\pi\hbar} \int_{-\infty}^{E_{F1}} dE = \frac{eE_{F1}}{\pi\hbar}. \tag{1}$$

A similar result is obtained for the right-left current,

$$J_- = 2 \cdot \frac{e}{2\pi\hbar} \int_0^\infty v dp = \frac{e}{\pi\hbar} \int_{-\infty}^{\mu_2} dE = \frac{eE_{F2}}{\pi\hbar}, \tag{2}$$

so, the net current from left to right is:

$$J = J_- - J_+ = \frac{e}{\pi\hbar}(E_{F1} - E_{F2}). \tag{3}$$

Then, width the substitution $E_{F1} - E_{F2} = eV$, we obtain,

$$J = \frac{e^2}{\pi\hbar} V. \tag{4}$$

The proportionality factor between current and voltage is the quantized conductance for a single band:

$$g_0 = \frac{e^2}{\pi\hbar}. \tag{5}$$

Assuming transport is occurring in N bands (channels) under the Fermi level, the total conductance is,

$$g = N \cdot g_0. \tag{6}$$

This expression clearly reveals that the conductance is quantized in units of g_0. In reality, there is a finite probability that in going from the reservoir into the wire, and vice versa, some electrons may be backscattered, in which case the number of bands through which transport is operative is less than N. In that case the effective value for N is conductance is given by:

$$N_{Effective} = \sum_{i1}^{N} T_n(E_F), \tag{7}$$

where T_n is the transmission coefficient of band n. Clearly, casting the conductance in terms of the transmission coefficient uncovers its dependence on the wave nature of the electron.

3.1.1.2 Quantum Point Contacts

In deriving the quantized electrical conductance of a quantum wire above it was pointed out that it is proportional to N, the number of bands through which transport is operative. The quantum point contact (QPC), Fig. 3-3, represents a virtually zero-length quantum wire, in which the details of T_n dominate transport and are made patently manifest in the conductance.

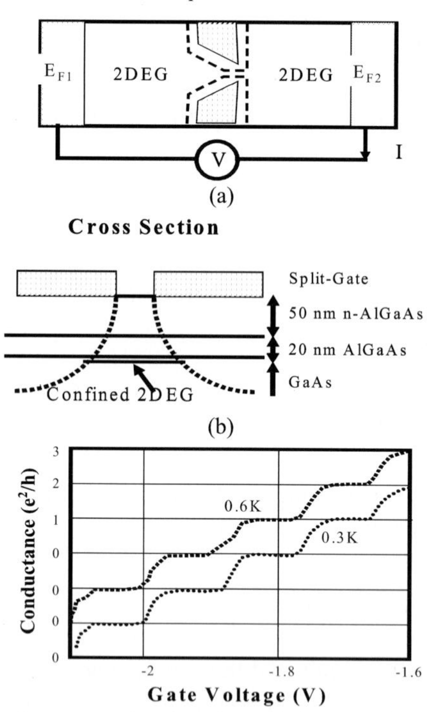

Figure 3-3. Quantum point contact. (a) Top view. (b) Cross-section. (c) Conductance versus gate voltage. (*After* [114].)

In the QPC a constriction is formed by modulating via, e.g., depletion regions, the width of the channel between two two-dimensional electron gas

(2DEG) regions, Figure 3-3 (a), (b). A rendition of the first experimental demonstration of the effect is shown in Figure 3-3(c). It is observed that the conductance decreases approximately linearly as the gate voltage is increased negatively, i.e., as the constriction or channel width narrows. In particular, at V_G=-2.2V, the channel is pinched-off and the conductance is zero. Notice also, that the conductance decreases in discrete steps of $2e^2/h$.

An explanation of the observed quantized conductance was attributed to the resistance of the constriction upon comparison with the semi-classical formula for the conductance of a constriction in a 2DEG, denoted G_S, after Sharvin who derived it [68]. G_S is given by,

$$G_S = \frac{e^2}{\pi}\frac{dN^{2D}}{dE}v_F W \ , \tag{8}$$

where $dN^{2D}/dE = m^*/\pi\hbar$ is the quantum mechanical density of states, including a factor of two for spin, $v_F = \hbar k_F/m^*$ is the Fermi velocity, with $k_F = 2\pi/\lambda_F = \sqrt{2\pi n_S}$ being the Fermi vector and n_S the 2DEG electron density, and W is the width of the constriction. Rewriting (65) so that the quantized conductance becomes explicit, one obtains,

$$G_S = \frac{2e^2}{h}\frac{k_F W}{\pi} = \frac{2e^2}{h}\frac{2W}{\lambda_F} . \tag{9}$$

The fact that this equation includes the ratio W/λ_F suggested that, experimentally, there should be deviations due to the manifestation of the wave nature of electrons whenever $\lambda_F \sim W$. In particular, it was determined that the plateau values of conductance are obtained whenever W is an integral multiple of $\lambda_F/2$. Therefore, the quantized conductance is a manifestation of the wave nature of electrons in that as the voltage is increased from pinch-off, a new mode (band) for transport becomes available every time the constriction widens by $\lambda_F/2$. The transmission coefficient of the constriction captures this [115]. The deviations from flatness of the conductance plateaus were attributed to scattering or to the abruptness of the constriction. Finally, as the temperature increases, the conductance steps smear out until at high temperature they disappear. This is due to the non-monoenergetic, wider, distribution of electrons launched by the reservoirs into the constriction [68] and exposes one of the practical limitations of QPCs, namely, that their utilization requires extremely low temperatures.

3.1.2 Quantum Resonant Tunneling

One of the fundamental devices exploiting the wave nature of electrons, and which finds practical application at room temperature, is the resonant tunneling diode (RTD) [116], [117], see Figure 3-4.

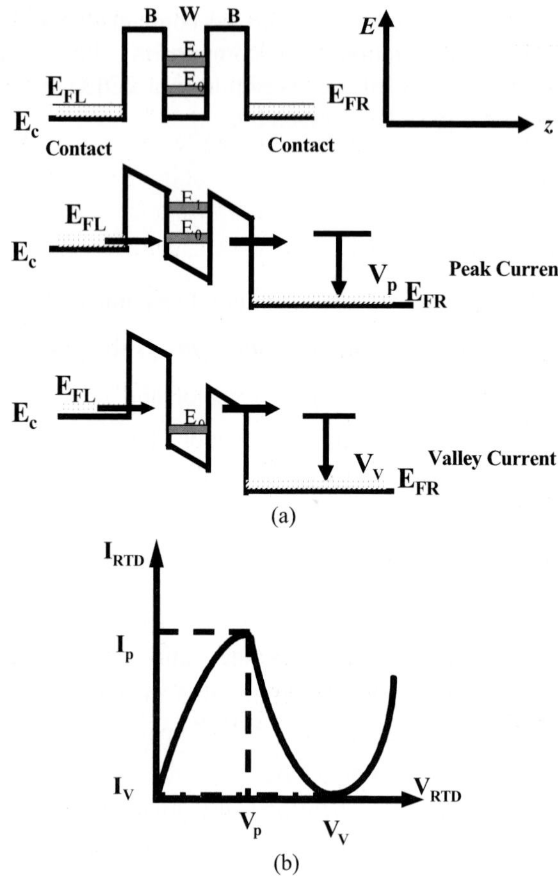

Figure 3.4. Resonant tunneling diode. (a) Energy band diagram and operation. (b) Current-voltage characteristic.

The RTD consists of a double barrier sandwiching a potential well, and in turn clad by two electron reservoirs (contacts). The potential well dimensions are of the order of tens of Angstrom, such that electrons in it are confined and, thus, can only exist in quantized energy levels. The barrier lengths are of the order of a few Angstroms, so that electrons can tunnel through them.

Resonant tunneling devices are implemented in a variety well/barrier materials systems [116], including, Type-I heterostructures (transport occurs exclusively in the conduction band) such as GaAs/$Al_xGa_{1-x}As$, InAs/AlSb, $In_{0.53}Ga_{0.47}As$/AlAs, and Type-II heterostructures (transport involves conduction and valence bands) such as GaSb/AlSb [118]

The ideal RTD current-voltage characteristic is shown in Fig. 3-4(b) and, with respect to Fig. 3-4(a), an accepted plausible explanation of it is as follows [116[, [117]. With no voltage applied, the system is in equilibrium as no forces are experienced by the electrons in the contacts and no current flows: (1) As the voltage is increased electrons tunnel the left-hand barrier, propagate through the well and tunnel through the right-hand barrier, and an increasingly large current flow; (2) When the voltage is such that the energy of the incoming electron distribution overlaps the first quantized energy of the well, E_0, maximum current transmission is achieved, this is the resonant tunneling condition; (3) When the overlap decreases, at higher applied voltages, the transmission, and thus current, rapidly decreases, thus the negative resistance region is produced. This explanation assumes the electron momentum transverse to the well is conserved.

Since the *intrinsic* time it takes an electron to traverse the structure is related to Heisenberg's uncertainty principle, $\tau = \hbar/\Gamma$, where Γ is the energy width of the quantized level, the process is very fast, i.e., ~1ps, so the devices are ideal for THz applications [118, 119].

The simulation and modeling of RTDs is a relatively mature subject [116-123] and includes a variety of approaches ranging from those neglecting scattering and charge effects to those including them to a variety of degrees. These models typically reproduce features of the I-V curve related to energy levels in the device, such as the voltages at which peak and valley currents occur, but not the magnitudes of these currents. A typical approach is the two-band tight-binding model, exposed by Schulman [124] for modeling a GaAs-GaAlAs RTD. In particular, by neglecting scattering and charge effects it focuses on calculating the transmission coefficient of the structure by employing an atom-to-atom transfer matrix technique that builds up the electron wave function as it propagates through the device layers. The model divides the structure as shown in Figure 3-5, assumes that the wave function is a combination of *s*-like orbitals on each cation (Ga, Al) and a *p*-like orbital on each anion (As), of the form,

$$\Psi = C_S \phi_S + C_P \phi_P, \tag{10}$$

and sets up a tight-binding Hamiltonian of the form,

$$H_I = \begin{bmatrix} E_S & U(1-e^{-ika/2}) \\ U(1-e^{-ika/2}) & E_p \end{bmatrix}, \tag{11}$$

$$H_I \Psi_I = E\Psi_I, \tag{12}$$

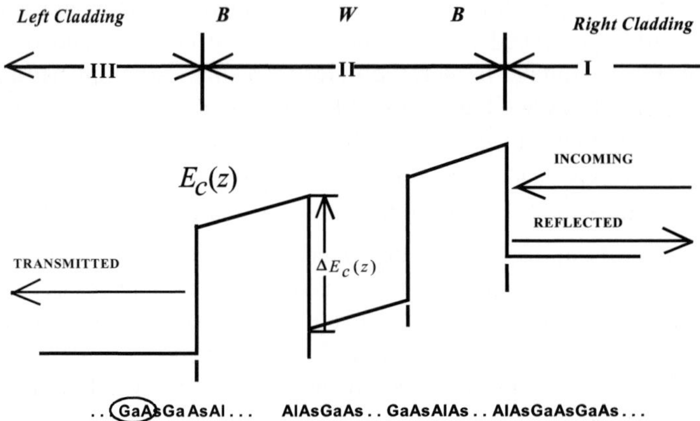

Figure 3-5. RTD structure for two-band tight-binding modeling.

where E_S and E_P are orbital energies prior to coupling to next neighbors, and a is the lattice constant. Next, solutions are formulated for the three regions as follows. For region I, we have (13) and (14).

$$\Psi_I^L = \frac{1}{\sqrt{2E - E_s - E_p}} (\sqrt{E - E_p}\phi_s + i\sqrt{E - E_s} e^{-i|k|(a/4)} \phi_p) \tag{13}$$

$$\Psi_I^R = \frac{1}{\sqrt{2E - E_s - E_p}} (\sqrt{E - E_p}\phi_s - i\sqrt{E - E_s} e^{-i|k|(a/4)} \phi_p) \tag{14}$$

For region II we have the transfer matrix (15).

$$\begin{bmatrix} C_s(n) \\ C_p(n) \end{bmatrix} = \begin{bmatrix} \dfrac{U(n-1,n-1)}{U(n-1,n)} & \dfrac{E_p(n-1)-E(n)}{U(n-1,n)} \\ \dfrac{[E-E_s(n)]U(n-1,n-1)}{U(n,n)U(n-1,n)} & \dfrac{U(n-1,n)}{U(n,n)} + \dfrac{[E-E_s(n)][E_p(n-1)-E]}{U(n,n)U(n-1,n)} \end{bmatrix} \begin{bmatrix} C_S(n-1) \\ C_P(n-1) \end{bmatrix} \tag{15}$$

For coupling regions II and III we have (16) and (17).

$$\Psi_{III} = C_s\phi_s + C_p\phi_p = C_L\Psi_{III}^L + C_R\Psi_{III}^R \tag{16}$$

$$\begin{bmatrix} C_L \\ C_R \end{bmatrix} = \sqrt{2E-E_s-E_p}\begin{bmatrix} \dfrac{1}{\sqrt{E-E_p}(1+e^{-ika/2})} & \dfrac{-i}{\sqrt{E-E_s}(e^{ika/4}+e^{-ika/4})} \\ \dfrac{1}{\sqrt{E-E_p}(1+e^{ika/2})} & \dfrac{i}{\sqrt{E-E_s}(e^{ika}+e^{-ika/4})} \end{bmatrix}\begin{bmatrix} C_s \\ C_p \end{bmatrix} \tag{17}$$

The dispersion relation, velocity, and overlap integral defining the tight-binding are given by (18), (19), and (20).

$$k(E) = \pm\left(\frac{4}{a}\right)\arcsin\left[\frac{1}{2U}\sqrt{(E-E_s)(E-E_p}\right] \tag{18}$$

$$v = \pm\frac{aU^2\sin(ka/2)}{\hbar(2E-E_s-E_p)} \tag{19}$$

$$U = \frac{2\hbar^2(E_s-E_p)}{m^*a^2} \tag{20}$$

Finally, the current is given by (21), where $x = E/kT$.

$$J(V) = \frac{e(kT)^2}{4\pi^2\hbar}\int_0^\infty dx\cdot 2\cdot\frac{v_I}{\left|C_L\right|^2 v_{III}}\cdot\frac{k_{III}}{\hbar v_{III}}\ln\left(\frac{1+e^{(E_F/kT-x)}}{1+e^{\left[(E_F-eV)/kT-x\right]}}\right) \tag{21}$$

This formulation, though not fully predictive, is a useful tool for the analysis and design of RTDs and related devices. A typical I-V curve produced using this formalism is shown in Figure 3-6.

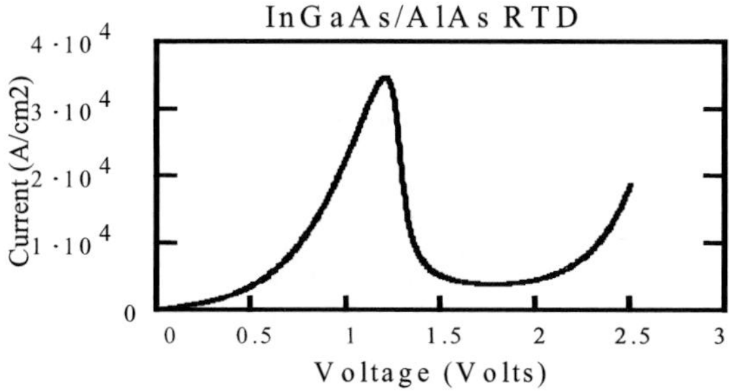

Figure 3-6. Current-voltage curve calculated via two-band tight-binding formalism.

3.1.3 Quantum Interference

While the RTD I-V characteristics are the result of constructive and destructive interference between the barriers, as a one-dimensional device these are really function of the degree of resonance with the energy levels in the well. When transport occurs in two dimensions, we may have constructive and destructive interference as a result of waves traveling thorough different paths that converge at one point.

3.1.3.1 Aharonov-Bohm Effect

The quintessential example of this type of interference, which also exposes the wave nature of electrons, is the Aharonov-Bohm (AB) effect [125], Figure 3-7. The essence of the AB effect, see Fig. 3-7, is that an electron beam, with wavefunction ψ_{in}, split at point A into two waves, ψ_1 and ψ_2, which subsequently follow paths ABF and ACF, around a solenoid establishing a magnetic flux ϕ_0 strictly in its interior, will gain respective phases S_1 and S_2 so that at F the wavefunction is,

$$\psi_F = \psi_1 e^{-iS_1/\hbar} + \psi_2 e^{-iS_2/\hbar}, \tag{25}$$

or, in other words, there is a phase difference $(S_1 - S_2)/\hbar$ between them. In particular, the phase shift is given by,

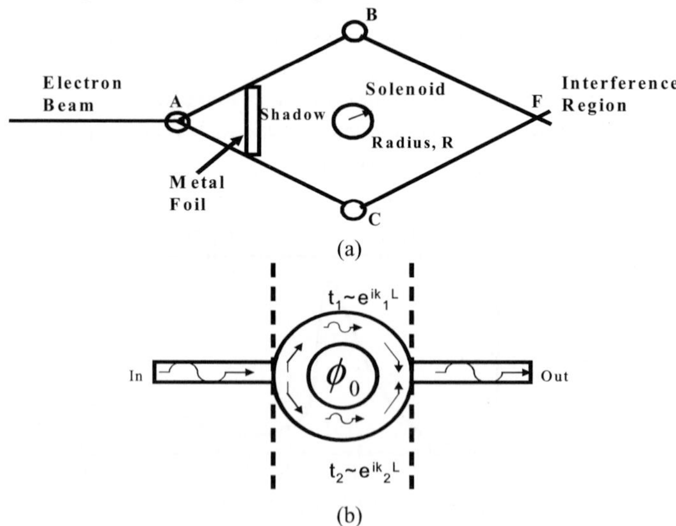

Figure 3-7. (a) Aharonov-Bohm-effect electron wave interference setup. (*After* [125].) (b) Sketch of metallic ring implementation.

$$\frac{\Delta S}{\hbar} = \frac{e}{\hbar}\int_{ABF}\left(\varphi dt - \frac{\vec{A}}{c}\cdot d\vec{l}\right) - \frac{e}{\hbar}\int_{ACF}\left(\varphi dt - \frac{\vec{A}}{c}\cdot d\vec{l}\right) = -\frac{e}{c\hbar}\oint \vec{A}\cdot d\vec{l}\,, \quad (26)$$

where φ is the scalar potential and $\vec{A}$ is the vector potential, which is related to the magnetic field inside the solenoid by (27).

$$\phi_0 = \oint \vec{A}\cdot d\vec{l} = \oint \vec{H}\cdot d\vec{s}\,, \quad (27)$$

The remarkable aspect of this effect is that, because of (27), it predicts, and has been confirmed, that a vector potential exists even where no magnetic field is existent, namely, outside the solenoid in this case, and this vector potential endows the wave functions with a phase shift difference which establishes that the electrons may exhibit interference. In particular, the phase shift may be expressed as,

$$\Delta\chi = \frac{e}{c\hbar}\phi_0\,, \quad (28)$$

so that when $\Delta\chi = 2\pi n$ there is constructive interference, and when $\Delta\chi = 2\pi(n+1/2)$ there is destructive interference.

3.1.4 Quantum Transport Theory

The wave nature of electrons is responsible for a number of phenomena, such as quantized electrical conductance, resonant tunneling, and quantum interference, which find their genesis in the quantum nature of electrons. Since, in fact, at dimensions approaching 100nm feature sizes, these effects are already beginning to dominate the characteristics of practical devices, the question of how to simulate the behavior of these quantum devices has received much attention. In this section, we focus on the principles of typical theoretical approaches to the quantum transport of heat and electrons.

3.1.4.1 Quantized Heat Flow

In bulk devices, the rate of heat conduction per unit area is proportional to the temperature gradient, i.e., Fourier's law, $\dot{Q}/A = -\kappa\nabla T$, where κ is the *bulk* coefficient of thermal conductivity. This expression assumes $\kappa = \gamma C v l_p$ [126], where γ is a numerical factor, C is the specific heat per unit volume, v is the velocity of sound, and l_p is the phonon mean free path, i.e., the typical device dimension $L >> l_p$. At nanoscale dimensions, however, $L < l_p$ and the phonons propagate *ballistically*. In this case, theory developed by Rego and Kirczenow [127], and experiments performed by Schwab, Henriksen, Worlock, and Roukes [128], have shown that the thermal conductance between isolated right and left temperature reservoirs, which are only interconnected through the device, is given by Landauer's theory as,

$$\kappa = \frac{1}{2\pi}\left\{\sum_{\alpha}^{N_\alpha}\int_0^\infty d\omega\hbar\omega\left(\frac{n_R(\omega)-n_L(\omega)}{\Delta T}\right)\zeta_\alpha(\omega)+\sum_{\alpha'}^{N_{\alpha'}}\int_{\omega_{\alpha'}(0)}^\infty d\omega\hbar\omega\left(\frac{n_R(\omega)-n_L(\omega)}{\Delta T}\right)\zeta_{\alpha'}(\omega)\right\}, \quad (29)$$

where $\omega_\alpha(k)$ and $\zeta_\alpha(k)$ are the frequency and phonon transmission probability of *normal mode* α, respectively, and $n_i(\omega) = \left(e^{\hbar\omega/k_B T_i} - 1\right)^{-1}$ represents the thermal distribution of phonons in reservoir with temperature T_i. While, it has been demonstrated in the works of Angelescu, Cross, and Roukes [129], and of Rego and Kirczenow [127], that the transmission probability is sensitive to the geometrical features of the nanoscopic systems, in particular, to phonon scattering due to surface roughness and transitions (non-adiabatic mode coupling), the main conclusion from (29) was that at low temperatures heat transport is mediated by a *universal constant*, namely, the *quantum of thermal conductance* due to phonons, $k_B^2\pi^2/3h$ [128]. This has serious implications pertaining to the maximum rate at which power can be dissipated in NanoMEMS, and indeed nanoscale thermal transport is a very active area of current research [130].

3.1.4.2 Fermi Liquids and Lüttinger Liquids

As suggested at the beginning of this chapter, transmission lines (TLs) are ubiquitous in circuits and systems at all length scales. Since TLs should simply transfer or guide signals from one location to another, without decreasing their amplitude or power, it is imperative that they exhibit the lowest possible loss. This is the reason why metals, due to their lowest resistivity, are preferably utilized to implement interconnects (TLs).

The resistivity of conventional (large-dimension) TLs reflects the dimensionality of electron motion. For instance, in TLs of rectangular cross-sectional area A, as dimensions shrink electron motion may become quantized in certain directions, thus giving rise the to the creation of energy sub-bands or "channels" in which transport can only occur once the electrons acquire the corresponding necessary energy, in other words, electrons behave as waves with discrete (quantized) wave vectors. The quantized electrical conductance is a manifestation of this. In contrast, electrons in TLs of relatively large dimensions may exist at virtually all energies and, if there were no interaction among electrons, they would behave as free particles. The theory of electron behavior in a metal, when electron-electron interactions are taken into account, is due to Landau [131] and is denoted *Fermi liquid theory*. A Fermi liquid is considered to be made up of "quasi-particles," which are fictitious entities that, while being physically different from electrons, behave similarly to electrons, but with a different mass and dispersion relationship.

When electron transport is confined along one dimension, a behavior different to that of free electrons and that of a Fermi liquid is observed. The new aggregate of entities is said to consist of another fictitious quasi-particle, namely, the plasmon, and is referred to as a *Lüttinger liquid (LL)*.

The distinction between Fermi liquid and Lüttinger liquid behaviors is important to the realization of nanoscale circuits and systems, not only from the point of view of TL properties, but also because their different behavior elicits new issues when connecting a Fermi liquid TL to a Lüttinger liquid TL. The fundamental aspects of Fermi and Lüttinger liquids are addressed next.

3.1.4.2.1 Fermi Gas

The Fermi liquid theory explains the success of the free-electron approximation in the calculation of transport problems, even in the context of electron-electron interactions. The usual point of departure for describing the Fermi liquid is the Fermi gas. This is the conceptual situation in which the metal is modeled as a solid of volume V and length L on a side $\left(V = L^3\right)$, which contains moving non-interacting electrons in much the same way as atoms and molecules move inside a gas container. Since the electrons are assumed to be independent, i.e., do not interact, they each obey a Schrödinger equation of the form [132],

$$H_0\psi = \left[\frac{p^2}{2m} + U(\vec{r})\right]\psi = E\psi, \tag{30}$$

where the potential energy is taken to be $U(\vec{r}) = 0$. The solution of this equation is then obtained by assuming that all space is filled by cubes of side

L, and that the wavefunction fulfills periodic boundary conditions at each of its faces, namely,

$$\psi(\vec{r}+L\hat{x})=\psi(\vec{r}+L\hat{y})=\psi(\vec{r}+L\hat{z})=\psi(\vec{r}), \tag{31}$$

These assumptions yield solutions of the form

$$\phi_{k\sigma}(\vec{r})=\frac{e^{i\vec{k}\cdot\vec{r}}}{\sqrt{V}}\chi_\sigma, \tag{32}$$

where $\sigma=\pm 1/2$ represents the two values of electron spin and χ_σ represents the two spin functions,

$$\chi_{1/2}=\begin{bmatrix}1\\0\end{bmatrix},\ \chi_{-1/2}=\begin{bmatrix}0\\1\end{bmatrix}. \tag{33}$$

Because of the periodic boundary condition, the wave vector is defined by,

$$k_x=\frac{2\pi}{L}n_x,\ k_y=\frac{2\pi}{L}n_y,\ k_z=\frac{2\pi}{L}n_z, \tag{36}$$

where $n_x,\ n_y,\ n_z=0,\pm1,\pm2,\ldots,$ $|k|^2=k_x^2+k_y^2+k_z^2$. The energy eigenvalues of (32) are given by,

$$E_{k\sigma}=E_k=\frac{\hbar^2k^2}{2m}. \tag{37}$$

The salient properties of the electron gas as a whole are captured by its wave function, its total energy, and various quantities such as its specific heat, and its magnetic susceptibility. The wave function is given by the Slater determinant [132],

$$\psi_{\nu_1\nu_2\nu_3\ldots\nu_N}(1,2,3,\ldots N)=\frac{1}{\sqrt{N!}}\begin{vmatrix}\phi_{\nu_1}(1) & \phi_{\nu_1}(2) & \ldots & \phi_{\nu_1}(N)\\ \phi_{\nu_2}(1) & \phi_{\nu_2}(2) & \ldots & \phi_{\nu_2}(N)\\ \ldots & \ldots & \ldots & \ldots\\ \phi_{\nu_N}(1) & \phi_{\nu_N}(2) & \ldots & \phi_{\nu_N}(N)\end{vmatrix}, \tag{38}$$

which ensures that the Pauli exclusion principle is obeyed, i.e., if two of the one-particle states ν_i are the same, then $\psi_{\nu_1\nu_2\cdots\nu_N}\equiv 0$. With $U(\vec{r})=0$, the lowest energy eigenvalue (ground state energy) is given by the sum of the one-electron energies up to a maximum energy level denoted by E_F and called Fermi energy. This is obtained when *N* electron states with energy less than E_F are occupied, and all states with energy greater than E_F are unoccupied. To obtain an expression for E_F, one pictures the states in (38) as a grid of point in $k_x k_y k_z$ space where they form a fine three-dimensional grid of spacing $2\pi/L$, such that a sphere centered at $\vec{k}=0$ would contain $\frac{4\pi}{3}k^3 \cdot \frac{1}{\left(\frac{2\pi}{L}\right)^3} = \frac{Vk^3}{6\pi^2}\cdot 2 = \frac{Vk^3}{3\pi^2}$ points of the grid when its radius is $\vec{k}$, including spin. Since each point in the grid represents one electron, the number of grid points contained in a sphere with the largest radius, k_F, corresponding to E_F must equal *N*,

$$\frac{Vk_F^3}{3\pi^2} = N. \tag{39}$$

Thus, the largest electron momentum is,

$$k_F = \sqrt[1/3]{\frac{3\pi^2 N}{V}} = \left(3\pi^2 n\right)^{1/3}, \tag{40}$$

where *n* is the electron density in the metal, and the Fermi energy is,

$$E_F = \frac{\hbar^2}{2m}\left(\frac{3\pi^2 N}{V}\right)^{2/3} = \frac{\hbar^2}{2m}\left(3\pi^2 n\right)^{1/3}. \tag{41}$$

At absolute zero, all levels are filled up to E_F. For an arbitrary energy *E*, less than E_F, the total number of electrons with energy less than *E* is given by,

$$N = \frac{V}{3\pi^2}\left(\frac{2m}{\hbar^2}\right)^{3/2}, \tag{42}$$

from where the density of states is given by,

$$D(E) \equiv \frac{dN}{dE} = \frac{V}{2\pi^2}\left(\frac{2m}{\hbar^2}\right)^{3/2}\sqrt{E}\,. \tag{43}$$

Excitation of the ground state of a Fermi gas requires, due to Pauli exclusion principle constraints, the addition of particles with momentum greater $|\vec{k}| > |k_F|$, or the destruction of a particles (creation of holes) with $|\vec{k}| < |k_F|$. However, if these particles came from outside the system, then the total number of particles N would change and we would have a different system. When one insists on inducing excitations that conserve the number of particles, then creating a particle with $|\vec{k}| > |k_F| = |\vec{k}|$, is accompanied by creating a hole with $|\vec{k}| < |k_F| = |\vec{k}'|$, i.e., particle-hole excitations which can be identified by two quantum numbers $\vec{k}, \vec{k}'$ are created.

These excitations may be caused by a number of influences, in particular, a rise in temperature or the application of a magnetic field k_F. Since, under no interaction, all states are occupied up to k_F, electrons closest to E_F will require the minimum energy to excite. Thus, the energy necessary to excite an electron of momentum k_1, for instance, is $E_{Excitation} = \frac{\hbar^2(k_F - k_1)^2}{2m}$. Temperature-induced excitations of the Fermi gas are captured by the specific heat, given by [28],

$$C_{el} = \frac{\partial E}{\partial T} = \frac{\pi^2}{3} D(E_F) k_B^2 T = \gamma T\,, \tag{44}$$

where k_B is Boltzmann's constant, and magnetic field-induced excitations are captured by the magnetic susceptibility given by,

$$\chi = \frac{M}{B} = 2D(E_F)\mu_B^2\,, \tag{45}$$

where μ_B is the Bohr magneton. Clearly, these quantities involve the density of states evaluated at one point, namely, the Fermi energy. This fact, coupled to the circumstance that, as long as one is dealing with a non-interacting free electron gas $D(E_F)$ will have the same value, suggests that solving both (44) and (45) for $D(E_F)$ and taking the ratio of the resulting quantity must be equal to one. This ratio, called the Wilson ratio, is given by [133],

$$R_W = \frac{\pi^2 k_B^2 \chi}{3\mu_B^2 \gamma}, \tag{46}$$

and captures the degree to which there are electron interaction effects. In particular, deviations from R_W signal the presence of interaction. Discussions on the Fermi liquid concept, which embodies phenomena due to electron-electron interation, usually make use of this index as a characterization parameter.

3.1.4.2.2 Fermi Liquids

Fermi liquid theory assumes that as the electron-electron interaction is turned on, from its zero value in the Fermi gas, the states in the now interacting system evolve directly from those of the noninteracting system, in such a way that the excited particles may also be labeled by momentum pairs $\vec{k}, \vec{k}'$, just as in the noninteracting electron case [134]. This circumstance is exemplified by the evolution of states in a noninteracting electron gas situated in an infinite-wall potential well as the interaction between them is turned on very slowly (adiabatically), see Figure 3-8 [134]. Having identical quantum labels for noninteracting electrons and quasi-particles implies that quantities that depend on these labels, such as the configurational entropy and the energy distribution, remain unchanged after the interaction is turned on [134]. Such is not case with the total energy because the energy of interaction modifies its value from the simple sum of that of the free particles.

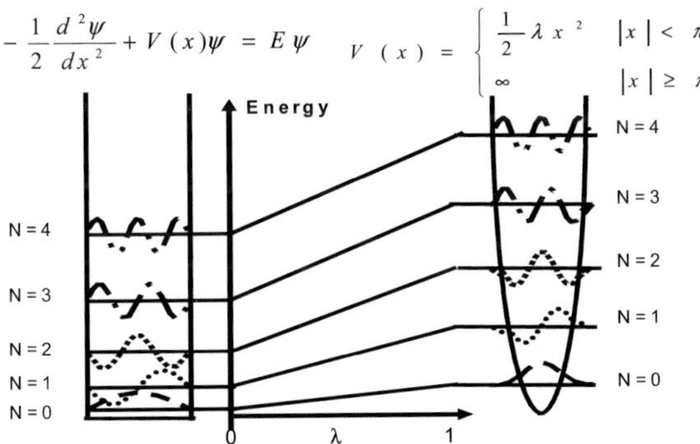

Figure 3-8. Adiabatic continuity explains how the labels of the energy states in a noninteracting electron gas may continue to be used as the interaction λ is turned on. Notice that, as the energy levels and their corresponding eigenfunctions evolve, the quantum labels (N) of the original noninteracting problem remain. *After* [134].

On the other hand, the excited particles, while finding themselves at the same $\vec{k}, \vec{k}'$ as the free elecrons, exhibit a different mass and a different E vs. $\vec{k}$ relationship than these, in particular, see Figure 3-9, interactions among the particles with states below E_F, and between these and the excited electrons with energy above E_F, are responsible for this. Thus, the dynamical properties of quasi-particles differ from those of free electrons. Under these circumstances, the theory assumes that for low-energy excitations, the quasi-particle distribution evolves in such a way that, if [133],

$$\begin{aligned} n_0(k) &= 1 \quad \text{if} \quad |\vec{k}| < k_F \\ &= 0 \quad \text{if} \quad |\vec{k}| > k_F \end{aligned} \tag{47}$$

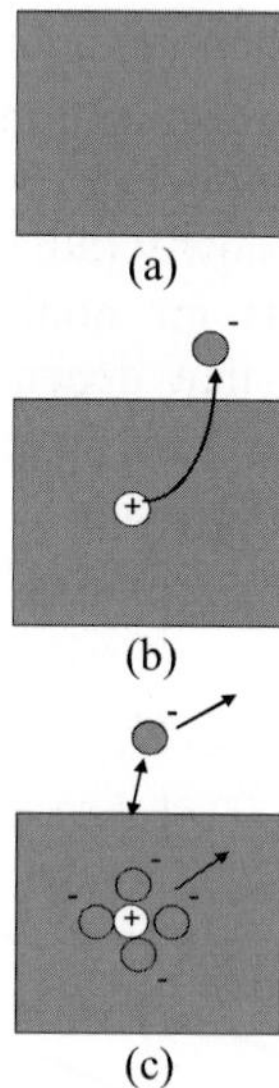

Figure 3-9. Fermi liquid representation. (a) Ground state. (b) Excited state. (c) The quasi-particle exhibits a new effective mass, m*, which derives from its interaction with ground state electrons as it moves through them. This effective mass is in addition to the mass derived from its interaction with the crystal lattice (captured by the energy band curvature), i.e., the dispersion relation *E vs. k*.

then the distribution of the noninteracting gas is $n_0(k)$, and, upon excitation $n_0(k) \rightarrow n_0(k) + \delta n(k)$, where $\delta n(k) = +1$ when a quasi-particle is excited, and $\delta n(k) = -1$ when a quasi-hole is excited. Here, $k = (\vec{k}, \sigma)$, and $\sigma = (\uparrow, \downarrow)$ represents spin. Similarly, the corresponding energy change is assumed to be given by,

$$\delta E = \sum_k E_k^0 \delta n(k) + \frac{1}{2\Omega} \sum_{kk'} f(k,k') \delta n(k) \delta n(k'), \tag{48}$$

where the first term represents the energy of an individual quasi-particle, defined as,

$$E_k = \frac{\hbar^2 k_F \left(|\vec{k}| - k_F \right)}{2m^*}, \tag{49}$$

with m^* representing its effective mass, and the second term, in particular, $f(k,k')$ capturing the interaction energy between quasi-particles. Further, in analogy with the the case of noninteracting states, the probability of a quasi-particle occuping a state k obeys Fermi statistics,

$$n(k) = \frac{1}{1 + e^{\beta E_k}}, \tag{50}$$

where $\beta = 1/k_B T$ and E_k is given by (49). In the case of the Fermi liquid, it has been found that calculations may be simplified by expressing the interaction function as the sum of symmetric and anti-symmettric terms, namely,

$$f(\vec{k}\uparrow, \vec{k}'\uparrow) = f^s(\vec{k}, \vec{k}') + f^a(\vec{k}, \vec{k}'), \tag{51a}$$

and

$$f(\vec{k}\uparrow, \vec{k}'\downarrow) = f^s(\vec{k}, \vec{k}') - f^a(\vec{k}, \vec{k}'). \tag{51b}$$

Then, assuming that these interaction functions exhibit rotational symmetry, and vary slowly with $|\vec{k}|$, the approximation $|\vec{k}| = |\vec{k}'| = k_F$ is made, which permits a Legendre expansion of the form [132],

$$f^{s,a}(\vec{k}, \vec{k}') = \sum_{L=0}^{\infty} f_L^{s,a} P_L(\cos\theta) \ , \cos\theta = \frac{\vec{k} \cdot \vec{k}'}{k_F^2}, \tag{52}$$

where P_L are the Legendre polynomials. Inversion of the expansion gives the coefficients,

$$f_L^{s,a} = \frac{2L+1}{4\pi} \int d^2\Omega P_L(\cos\theta) f^{s,a}(\vec{k},\vec{k}'), \tag{53}$$

which in normalized form are rewritten as,

$$F_L^{a,s} = \frac{k_F m^*}{\pi^2} f_L^{a,s}. \tag{54}$$

Following the considerations in the discussion of the noninteracting electron gas, excitations of the Fermi liquid are also captured by the specific heat and the magnetic susceptibility. These calculations assume that, for low energies, $E_k \to E_k^0$ and $m \to m^*$, and yield [133],

$$\gamma = \frac{m^* k_F k_B^2}{3}, \tag{55}$$

and

$$\chi = \frac{1}{1+F_0^a} \cdot \frac{\mu_B^2 k_F m^*}{\pi^2} \tag{56}$$

from where the Wilson ratio is given by (57) in terms of the Landau parameter F_0^a.

$$R_W = \frac{1}{1+F_0^a}. \tag{57}$$

For the quintessential example of a Fermi liquid, namely, liquid helium 3 (^3He), a coefficient of $F_0^a \approx -0.7$ [133] was obtained experimentally, resulting in a Wilson ratio $R_W \approx 3.33$, which denotes strong interaction.

Landau's Fermi liquid theory succeeds in capturing the phenomenology of near equilibrium properties, as shown above, however, in situations when it is not possible to write a simple expansion for *f*, as is the case in highly anisotropic metals, the application of the theory to obtain quantitative results becomes impossible [133], [134].

A more fundamental limitation of the theory derives from the circumstances under which the concept of quasi-particles is valid, namely, when their lifetime is longer than the time it takes to turn on the interaction [133], [134]. In particular, if the Hamiltonian for the interacting system as a function of the interaction parameter V and turn-on time τ_V is given by [135],

$$H = H_0 + Ve^{t/\tau_V}\ , \tag{58}$$

then the time it takes a quasiparticle of excitation energy ε to decay, τ_ε, must be much greater than the interaction turn-on time, τ_V, and also much greater than the time it takes the quasiparticle of to absorb the excitation energy, given by Heisenberg's uncertainty principle $\varepsilon/\hbar$ form,

$$\tau_\varepsilon >> \tau_V >> \frac{\hbar}{\varepsilon}. \tag{59}$$

Obviously, at large excitation energies $\Delta E = \varepsilon$, the associated time during which this energy is absorbed $\hbar/\varepsilon$ may become much smaller than the lifetime τ_ε, which means that no quasiparticle has a chance to form and, thus, the quasiparticle concept breaks down. An estimate of this lifetime is given in [134] by calculating the decay rate of a quasi-particle with energy ε above the Fermi energy E_F, at absolute zero. Using Fermi's golden rule, which describes the transition between initial states i and final states f elicited by a scattering potential V_{if},

$$\frac{1}{\tau_\varepsilon} = \frac{2\pi}{\hbar}\sum_f \left|V_{if}\right|^2 \delta\left(\varepsilon - \varepsilon_f\right), \tag{60}$$

assuming V_{if} is constant and enforcing conservation of energy and Pauli exclusion principles, see Figure 3-10, one obtains,

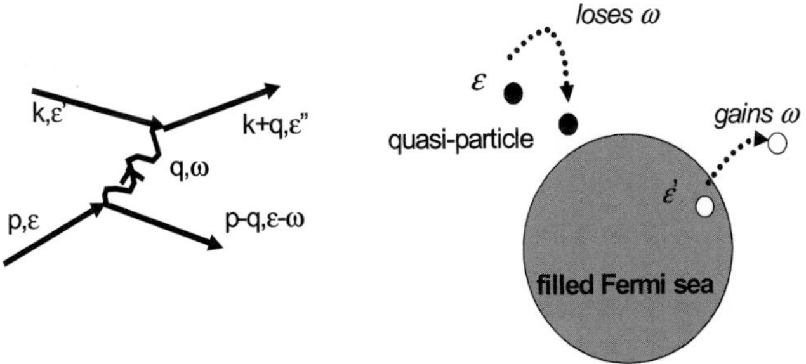

Figure 3-10. Energy relationship of quasi-particle scattering process. The energy ω lost in a scattering event by the quasi-particle must be lower than its initial energy ε, and there must be an electron state at an energy ε' capable of absorbing at this energy ω.

$$\begin{aligned}\frac{1}{\tau_\varepsilon} &\sim \frac{2\pi}{\hbar}|V|^2 \int_0^\varepsilon D(E_F)d\omega \int_0^\omega D(E_F)d\varepsilon' \int_{-\infty}^{\infty} \delta(\varepsilon-\omega-\varepsilon'+\varepsilon'')D(E_F)d\varepsilon'' \\ &\sim \frac{\pi}{\hbar}|V|^2 D^3(E_F)\varepsilon^2\end{aligned} \quad . \quad (61)$$

This results suggests that, the smaller the quasi-particle (excitation) energy ε, the longer will its lifetime be, in particular, as $\varepsilon \to 0$, the lifetime tends to infinity. An interesting result that relates the validity of the quasiparticle concept to the dimensionality d of the system was derived by Schofield [134], by making a change of variables to express (61) in terms of the momentum and energy transferred. His result was the expression,

$$\frac{1}{\tau_\varepsilon} = \frac{2\pi}{\hbar} \int_0^\varepsilon D(E_F)\omega d\omega \int_{\frac{\omega}{\hbar v_F}}^{2k_F} \frac{q^{d-1}dq}{(2\pi/L)^d} \frac{|D(q,\omega)|^2}{(\hbar v_F q)^2} \quad . \quad (62)$$

The integral (62) is interpreted as follows [136]: 1) The integral over ω accounts for the number of possible hole excitations that can be created; 2) The lower limit of the momentum integral, over q, signifies that a minimum momentum must be transferred to give a change in energy of ω; 3) The denominator $(\hbar v_F q)^2$ in the integrand embodies the fact of already having performed integration over the direction of the momentum and it reflects that there is an increased time available for small deflections; 4) The numerator, $D(q,\omega)$ is the matrix element for the scattering process. Examination of the impact of setting the dimension to $d = 1$ reveals that, if one assumes $D(q,\omega)$ to be constant, then due to the singularity of the q integral, the projected quasiparticle lifetime τ_ε, is not much greater than, but in fact is it close to, $\hbar/\varepsilon$. Therefore, (62) is violated as the quasiparticle, in principle, can never have enough time to form. The importance of this result is that *Fermi liquid theory breaks down when applied to one-dimensional metallic systems*, such as are typical at nanoscales. The new situation is described by the concept of the Lüttinger liquid.

3.1.4.2.3 Lüttinger Liquids

The term *Lüttinger liquid* is used to denote the behavior of interacting electrons confined to one-dimensional transport [137]. Such behavior is unraveled by solving the interacting electron problem. The Hamiltonian in question, given by (see Appendix B),

$$H_0 = \frac{2\pi}{L}\frac{v}{2}\left\{\left(\frac{1}{g}+g\right)\sum_{v=L,R}\left[\frac{1}{2}\hat{N}_v^2+\sum_q n_q b_{qv}^+ b_{qv}\right]\right. \\ \left. +\left(\frac{1}{g}-g\right)\left[\hat{N}_L\hat{N}_R-\sum_q n_q\left(b_{qR}b_{qL}+b_{qR}^+b_{qL}^+\right)\right]\right\}, \quad (63)$$

must be diagonalized to determine the pertinent types of solutions holding in one dimension. This Hamiltonian diagonalization is facilitated by the procedure of *bosonization* [137]-[139] discussed in detail in Appendix B. In essence, one-dimensional bosonization transforms a nondiagonal fermionic Hamiltonian into a diagonal bosonic one, with the assumption that the one-dimensional dispersion relatonship is linear, and given by $E(k) = v_F\hbar|k-k_F|$ [134]. The nature of this dispersion relation gives rise to the transport characterization in terms of *spinless* left- and right-moving electrons with respective electron densities N_L and N_R, the parameter *g*, which captures the electron-electron interaction strength in the problem, and the Fermi velocity $\mathrm{v_F}$. Kane and Fisher [140] have captured this phenomenology with he following set of expressions. The Hamiltonian (63) is rewritten as [140], [141],

$$H_0 = \pi v_0\left[N_R^2 + N_L^2 + 2\lambda N_R N_L\right], \quad (64)$$

with

$$\mathrm{v}_0 = \mathrm{v}\left[\frac{\left(g^{-1}+g\right)}{2}\right], \qquad \lambda = \frac{\left(1-g^2\right)}{1+g^2}, \quad (65)$$

with λ as the interaction strength parameter between the left- and right-moving electron species, and *g*, called the Lüttinger parameter. For $g=1$ the interaction is zero, and the Hamiltonian then captures the behavior of a noninteracting electron gas with velocity equal to the Fermi velocity $\mathrm{v}_0 = \mathrm{v}_F$. From (65) it is seen that repulsive interactions, which per (64) imply $\lambda > 0$, lead to $g < 0$, and the opposite is true for attractive interactions. In terms of the two-particle interaction potentials, V_2 and V_4, between fermions moving in opposite directions, namely, left and right, and either both left- or both right-moving, respectively, v and g are given by,

$$\mathrm{v} = \mathrm{v}_F \sqrt{1 + \frac{V_4}{\pi \mathrm{v}_F} + \frac{V_4^2 - V_2^2}{4\pi^2 \mathrm{v}_F^2}}, \tag{66}$$

and

$$g = \sqrt{\frac{1 + V_4/2\pi \mathrm{v}_F - V_2/2\pi \mathrm{v}_F}{1 + V_4/2\pi \mathrm{v}_F + V_2/2\pi \mathrm{v}_F}}. \tag{67}$$

Kane and Fisher [140] interpret v in the case $V_4 = V_2 > 0$, as the plasmon velocity, which increases above v_F when the repulsive interactions reduce the compressibility of the electron gas.

When the electron spin is included in the Hamiltonian, the interaction becomes,

$$V(x - x')\tilde{\rho}(x)\tilde{\rho}(x') \Rightarrow V_{\sigma\sigma'}(x - x')\tilde{\rho}_\sigma(x)\tilde{\rho}_{\sigma'}(x'). \tag{68}$$

In this case, the kinetic energy part of the Hamiltonian may be written as follows [133].

$$\begin{aligned} H_{kin} &= \mathrm{v}_F \sum_{k,s} \left((k - k_F) c^+_{+,k,s} c_{+,k,s} + (-k - k_F) c^+_{-,k,s} c_{-,k,s} \right) \\ &= \frac{2\pi \mathrm{v}_F}{L} \sum_{q>0,\alpha=\pm,s} \rho_{\alpha,s}(q) \rho_{\alpha,s}(-q) \end{aligned}, \tag{69}$$

where the substitution,

$$\rho_{\pm,s}(q) = \sum_k c^+_{+,k,s} c_{+,k,s}, \tag{70}$$

representing density operators for spin projections $s = \uparrow, \downarrow$ has been made. The potential energy, in turn, contains two types of interaction, namely, backward scattering and forward scattering. The backward scattering Hamiltonian is given by,

$$H_{\mathrm{int\_1}} = \frac{1}{L} \sum_{k,p,q,s,t} g_1 c^+_{+,k,s} c_{+,k,t} c_{+,p+2k_F+q,t} c_{-,k-2k_F-q,s}, \tag{71}$$

which captures scattering events in which $(k_F, s; -k_F, t) \rightarrow (-k_F, s; k_F, t)$ for $s \neq t$. The forward scattering Hamiltonian is given by,

$$H_{\text{int\_2}} = \frac{1}{L}\sum\left(g_2(q)\rho_{\alpha,s}(q)\rho_{-\alpha,t}(-q) + g_4(q)\rho_{\alpha,s}(q)\rho_{\alpha,t}(-q)\right). \quad (72)$$

The full bosonized Hamiltonian has been shown by Schulz [133] to take the form,

$$\begin{aligned} H &= H_{kin} + H_{\text{int\_1}} + H_{\text{int\_2}} \\ &= H_\rho + H_\sigma + \frac{2g_1}{(2\pi a)^2}\int dx \cos\left(\sqrt{8}\phi_\sigma\right), \end{aligned} \quad (73)$$

where a is a short-distance cutoff, and for $\nu = \rho, \sigma$,

$$H_\nu = \int dx \left(\frac{\pi u_\nu K_\nu}{2} \Pi_\nu^2 + \frac{u_\nu}{2\pi K_\nu} (\partial_x \phi_\nu)^2 \right), \quad (74)$$

with,

$$u_\nu = \sqrt{\left(\mathrm{v}_F + \frac{g_{4,\nu}}{\pi}\right)^2 - \left(\frac{g_\nu}{2\pi}\right)^2}, \quad K_\nu = \sqrt{\frac{2\pi \mathrm{v}_F + 2g_{4,\nu} + g_\nu}{2\pi \mathrm{v}_F + 2g_{4,\nu} - g_\nu}}, \quad (75)$$

and $g_\rho = g_1 - 2g_2$, $g_\sigma = g_1$, $g_{4,\sigma} = 0$.

Schulz [133] has exposed a number of situations by examining (75). For instance, he points out that a noninteracting system, for which $u_\nu = \mathrm{v}_F$ and, thus exhibits equal charge and spin velocities, is obtained by setting $K_\nu = 1$. That if $g_1 = 0$, then there is no backscattering and (75) describes uncoupled charge and spin density oscillations with a dispersion relation $\omega_\nu(k) = u_\nu |k|$ and the system is conducting.

The Hamiltonian (75) offers, as one of its consequences, the possibility of complete separation in the dynamics of spin and charge. In particular, if $u_\rho \neq u_\sigma$, then spin and charge waves propagate with different velocities. The electron, in this case, is said to dissolve into two particles, namely, a spin particle, called a *spinon*, and a charge particle, called a *holon* [134]. A simple picture for visualizing spin-charge separation is shown in Figure 3-11.

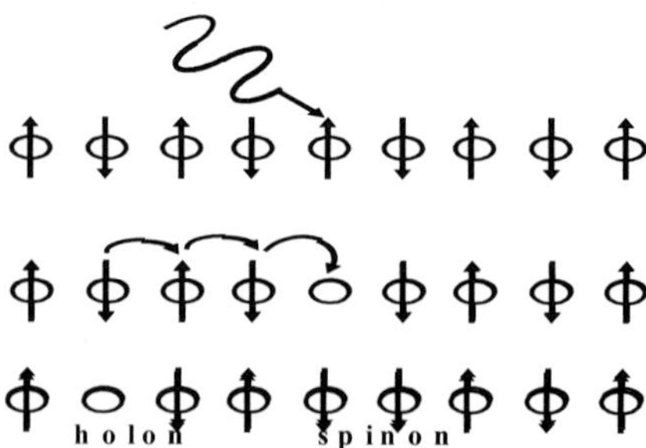

Figure 3-11. Illustration of spin charge separation. If a photon impinges on an antiferromagnetic Mott insulator an removes an electron, the disruption left behind changes both the spin and charge order. Electron motion into the vacant site results in spin and charge separation, giving rise to two distinct particles, namely, a *holon* and a *spinon*. (*After* [134].)

Qualitatively, the pertinent physics of the Lüttinger liquid follow from the dispersion relation and may be surmised from Fig. 3-11 [134]. An examination of this figure indicates that, due to the linear dispersion relation, changes in momentum determine energy changes. In particular, a momentun excitation $\vec{q}$ imposed on the 1D electron system, will cause a compression and rarefaction of the electron density with a wavelength $2\pi/|\vec{q}|$. The degrees of compression and rarefaction embody a density wave, and has two consequences. First, because $\vec{q}$ determines the kinetic energy E in a unique way, the density wave has a well-defined kinetic energy. Second, the concomitant density will depend on both the spin interaction and the Coulomb interaction amongst electrons which, being functions of distance, embody the potential energy of the system. Therefore, the total energy of the system may be specified by the properties of a density wave. This density wave, in turn, contains a spin density and a charge density. This spin-charge separation and coexistence is the hallmark of the Lüttinger liquid.

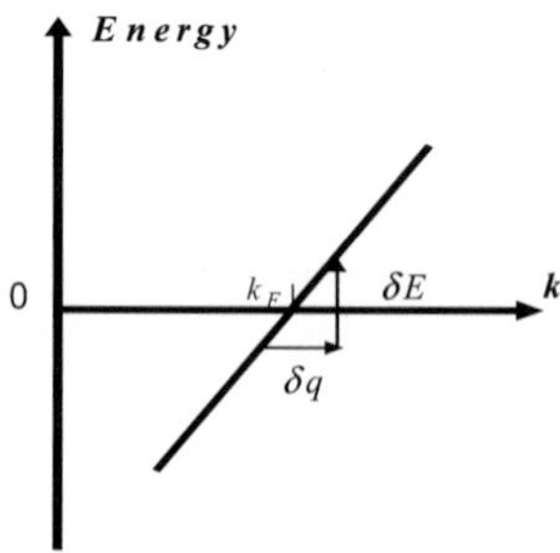

Figure 3-12. Excitation of electron-hole pairs in one-dimensional structure. *After* [134].

The behavior of the Lüttinger liquid at low energy excitations is captured by the specific heat and the magnetic susceptibility. The specific heat is given by,

$$\gamma/\gamma_0 = \frac{1}{2}\left(\frac{v_F}{u_\rho} + \frac{v_F}{u_\sigma}\right), \tag{76}$$

where γ_0 is the specific heat coefficient for noninteracting electrons at Fermi velocity v_F, and the spin susceptibility is given by,

$\chi/\chi_0 = \frac{v_F}{u_\sigma}$. The Wilson ratio is given by [133],

$$R_W = \frac{\chi}{\gamma}\frac{\gamma_0}{\chi_0} = \frac{2u_\rho}{u_\rho + u_\sigma}. \tag{77}$$

The presentation in this section has exposed the fact that in one-dimensional transport, the quasi-particles of a Fermi liquid morph into two new entities, namely, spinons and holons, which, individually, transport spin and charge, respectively, and characterize the Lüttinger liquid. It will be seen in the next section, that the manifestation of spin-charge separation is responsible for a quantitative change in the behavior of 1D TLs.

3.2 Wave Behavior in Periodic and Aperiodic Media

The ability to create patterns of very high precision, made available by NanoMEMS fabrication technology, will endow engineers with the ability to effect signal processing on a variety of wave phenomena, e.g., electronic, electromagnetic, acoustic, etc. Much of this functionality will exploit the phenomenon of band gaps; typically, domains of energies or frequencies in which wave propagation is forbidden. In what follows, the topics of electronic [28] and photonic bandgaps [51, 142], are addressed.

3.2.1 Electronic Band-Gap Crystals

3.2.1.1 Carbon Nanotubes

Carbon nanotubes (CNTs) were already introduced in Chapter 1. They are a relatively new type of material and are considered by many to be the quintessential nanotechnology device. Their properties are related to those of a 2D perodic graphite sheet, see Figs. 3-13.

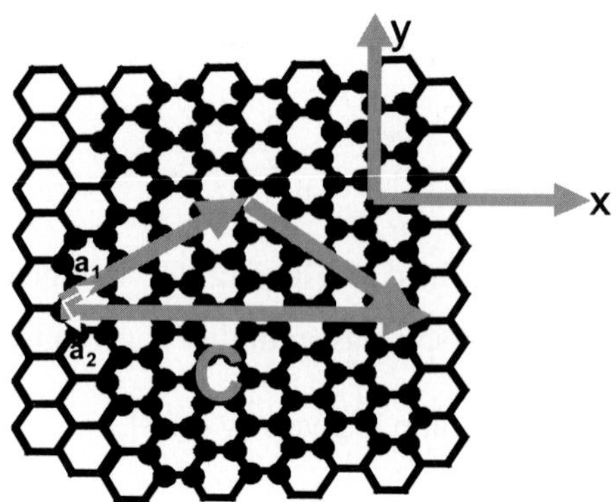

Figure 3-13. Sketch of a graphene lattice, a single sheet of carbon atoms arranged in the honeycomb structure, showing vectors utilized in describing the lattice. In this case, the vector **C** is defined by the pair n=4, m=4, i.e., (4, 4).

The graphene lattice is defined by a vector **C** of the form $\mathbf{C} = n\mathbf{a}_1 + m\mathbf{a}_2$, where $\mathbf{a}_1$ and $\mathbf{a}_2$ are the unit cell base vectors of the graphene sheet, Fig. 3-13, with $|a_1| = |a_2| = 0.246\text{nm}$. The pair of integers (*n*, *m*), where $n \geq m$, is used to represent a possible CNT structure [46]. Three types of CNT structures are typically identified according to how the conceptual graphene rolling into a cylinder is effected, namely, the armchair, the zigzag, and the chiral CNT structures, see Fig. 3-14 [143]. The *chiral* angle, θ, of the wrapping vectors describing these CNTs are related to the indices *n* and *m* by the equation [46],

$$\theta = \sin^{-1} \frac{\sqrt{3}m}{2\sqrt{n^2 + nm + m^2}} \tag{78}$$

with $\theta = 0$ for the Zigzag CNT, $\theta = 30^\circ$ for the Armchair CNT, and $0 < \theta < 30^\circ$ for the Chiral CNT. The corresponding CNT diameter is given by,

$$d_{CNT}(\text{Å}) = 0.783\sqrt{n^2 + nm + m^2}\,. \tag{79}$$

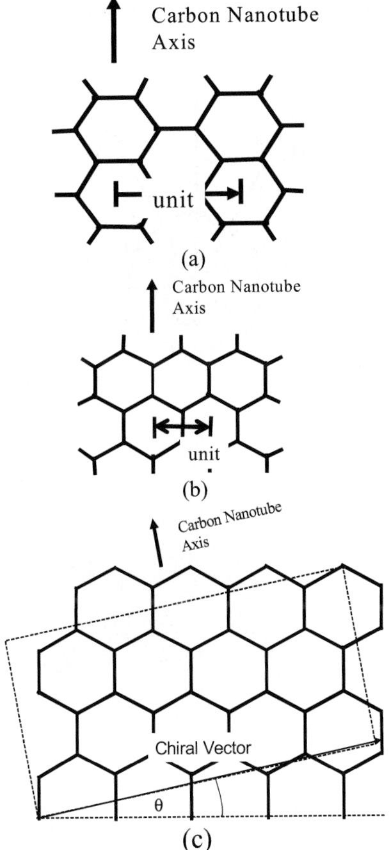

Figure 3-14. Carbon nanotube structures according to how the grapheme sheet is "wrapped." (a) Armchair. (b) Zig-zag. (c) Chiral. *After* [143].

As with conventional crystals, (electron) wave propagation is a function of the atomic (direct lattice) periodicity and its reciprocal lattice, and is captured by the dispersion relation, *E vs. k*. In the case of graphene, the direct lattice is of the honeycomb type, Fig. 3-15(a) and applying the tight-binding or linear combination of atomic orbitals (LCAO) method [64], the graphene band structure is obtained as,

$$E(\vec{k}) = \pm\gamma_1\sqrt{3 + 4\cos\left(\frac{\sqrt{3}ak_y}{2}\right)\cos\left(\frac{ak_x}{2}\right) + 4\cos\left(\frac{ak_x}{2}\right)}, \tag{80}$$

where a is the lattice constant, i.e., $a = \sqrt{3}a_0$. A plot of this function is shown in Figure 3-15(b). It may be noticed from this figure that at the K-points (the corners of the first Brilloin zone) there is zero gap between conduction and valence bands in graphene.

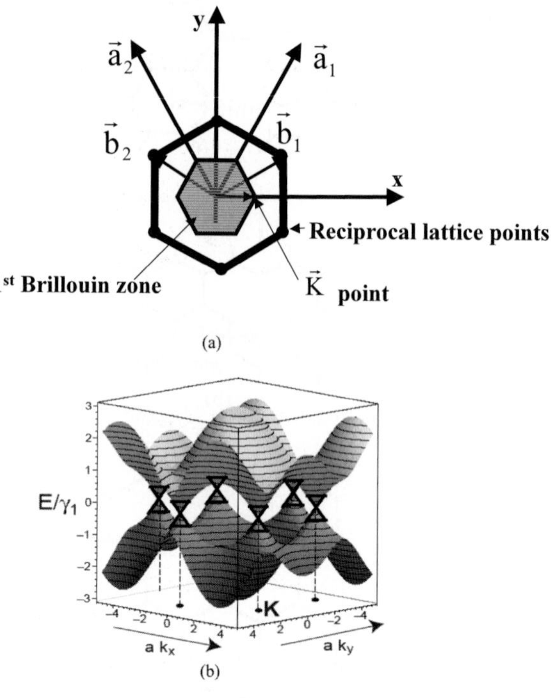

Figure 3-15 (a) Reciprocal lattice of graphene with the 1st Brilloin zone (shaded). $\vec{b}_1$ and $\vec{b}_2$ are the primitive lattice vectors. The K point lies at the edge of the BZ. 2D grapheme sheets "rolled" around the y axis, will give rise to armchair CNTs. (b) LCAO bandstructure of grapheme. The Fermi level lies at E=0. Courtesy of Prof. Christian Schönenberger, University of Basel, Switzerland].

The effect of rolling the graphene sheet to form the CNT manifests itself in the band structure as follows. On the one hand, the momentum of electrons along the circumference of the cylinder becomes quantized. On the other, propagation is now only possible along the cylinder axis, i.e., in one dimension, thus the concomitant CNT band structure corresponds to slices of the 2D graphene structure. When the slice passes through a K-point, the CNT is metallic since, at these points, the gap is zero; when it doesn't, it is semiconducting. In particular, CNT structure type and its electronic properties are related as follows [46]. For armchair CNTs, the circumferential momentum vector is quantized according to,

$$k_x^\nu = \frac{\nu}{N_x}\frac{2\pi}{\sqrt{3}\mathbf{a}}, \tag{81}$$

for $\nu = 1,\ldots,N_x$, where N_x is the number of unit cells spanning the circumference. Thus, it can be shown that an armchair CNT rolled such that its circumference lies along k_x and the transport longitudinal axis is along k_y, would have longitudinal 1D band structures at each of the discrete values of k_x given by (81), see Figure 3-18. Similarly, a zigzag CNT has its circumferential momentum vector quantized according to,

$$k_y^\nu = \frac{\nu}{N_y}\frac{2\pi}{\mathbf{a}}, \tag{82}$$

for $\nu = 1,\ldots,N_y$. In this case, the resulting CNT may be either metallic or semiconducting. Metallic, when its index n is divisible by three, in which case a slice passes through a K-point and the tube behaves as a 1D metal with Fermi velocity $v_F = 8\times10^5\ \mathrm{m/s}$ [144], and otherwise, semiconducting. In the context of ballistic CNTs, their conductance is given by Landauer's formula, $G = \left(Ne^2/h\right)T$, where N, the number of one-dimensional channels is four, due to electron spin degeneracy and the two bands at K- and K'-points, see Fig. 3-17(a). This works out to $G = \left(4e^2/h\right) = 1/6.5\mathrm{k}\Omega$, assuming T=1. The energy gap of semiconducting CNTs is related to their diameter by [144], [145],

$$E_{GAP} = \frac{4\hbar v_F}{3d_{CNT}} \approx \frac{0.9\mathrm{eV}}{d_{CNT}\,[nm]}. \tag{83}$$

In the general case of a chiral CNT, Dresselhaus *et al.* [146], [147] have shown that a metallic CNT is obtained whenever,

$$n - m = 3q, \tag{84}$$

where q is an integer. In summary, the current knowledge of electronic-structural properties of SWNTs is as follows [46]: all armchair tubes are expected to be metallic, one-third of zigzag and chiral tubes are expected to be metallic, and the rest are expected to be semiconducting [46].

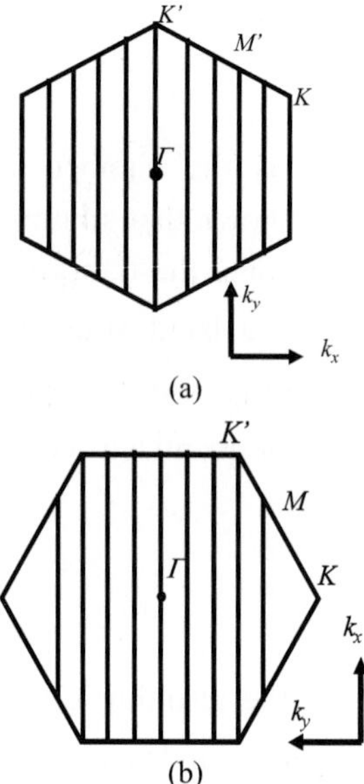

Figure 3-16 (a) A 1D band structure lies at each of the discrete values of k_x for a (5, 5) armchair CNT, dictated by the circumferential quantization in this direction. The armchair CNT is metallic. (b) A 1D band structure lies at each of the discrete values of k_y for a (9, 0) zigzag CNT, dictated by the circumferential quantization in this direction. The zigzag CNT is semiconducting. (*After* [46].)

It may be surmised from the slice passing through the K', K points, see Fig. 3.17, that each channel is four-fold degenerate, on account of spin degeneracy and the sublattice degeneracy of electrons in graphene [144].

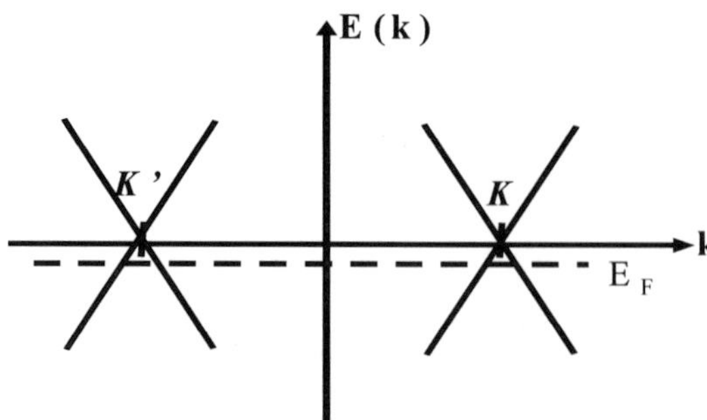

Figure 3-17. Energy band diagram of metallic CNT for slice through Fermi points K', K.

This fact has been utilized by Burke [148] to propose an AC circuit model for CNTs, including electron-electron interaction, see Fig. 3.18.

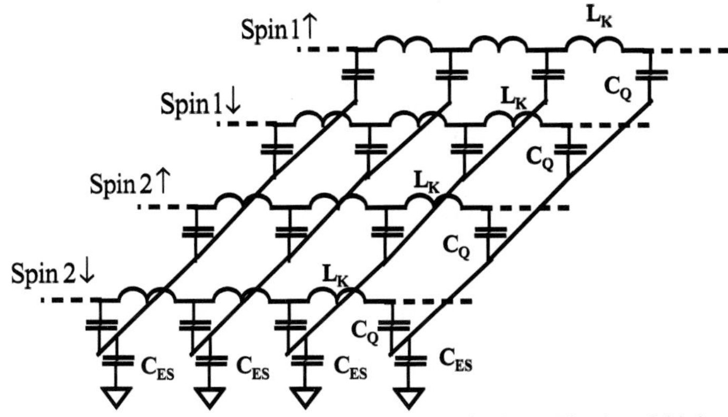

Figure 3-18. AC circuit model for interacting electrons in CNT. The four-fold degeneracy is captured by four channels. (*After* [148].)

The circuit model is interpreted by Burke [148] as follows. The circuit captures the existence of four modes, namely, *three spin modes*, which corresponds to a differential excitation, and *one charge mode*, which corresponds to common mode excitation. In the latter case (charge mode), all four transmission lines appear in "parallel", and they are characterized by an effective line possessing a charge-mode propagation velocity and characteristic impedance given by [148],

$$v_p = \sqrt{\frac{1}{L_K}\left(\frac{1}{C_Q} + \frac{4}{C_{ES}}\right)} = v_F\sqrt{1 + \frac{4C_Q}{C_{ES}}} \equiv \frac{v_F}{g}, \tag{85}$$

and,

$$Z_{c,CM} = \sqrt{\frac{4L_K}{C_{ES}} + \frac{L_K}{C_Q}} = \frac{1}{g}\frac{h}{2e^2}, \tag{86}$$

where, $L_K = h/2e^2 v_F$ (h is Planck's constant) is the kinetic inductance per unit length, $C_Q = 2e^2/hv_F$ is the quantum capacitance,and $C_{ES} = 2\pi\varepsilon/\cosh^{-1}(2h/d)$ (h here is the CNT-to-ground distance) is the electrostatic capacitance (the CNT-to-ground capacitance). Typical values for these parameters are: $L_K = 16\text{nH}/\mu\text{m}$, $C_{ES} = 50\text{aF}/\mu\text{m}$, and $C_Q = 100\text{aF}/\mu\text{m}$. The characteristic impedance for the three spin modes is given by,

$$Z_{c,DM} = \sqrt{\frac{L_K}{C_Q}} = \frac{h}{2e^2}, \tag{87}$$

which Burke interprets as defining the ratio of excitation voltage to elicited current when the spin wave is excited.

With diameters of the order of approximately 1*nm*, CNTs are ideal systems where the characteristics of Lüttinger liquids, namely, strong electron-electron interaction and spin-charge separation, should be manifest. Accordingly, efforts have been expended to develop ways of characterizing and ascertaining such behavior. Noticeable among these, is experimental work by Bockrath *et al.* [149] who deduced, from the measured 3D-1D tunneling conductance $dI/dV \propto V^{\alpha}$, CNT Lüttinger parameters g with values between 0.2 and 0.3. These were extracted from comparison of measurement to the theoretical relations $\alpha_{\mathbf{End}} = (g^{-1} - 1)/4$ or $\alpha_{\mathbf{Bulk}} = (g^{-1} + g - 2)/8$, for 3D-1D contacts located at the end or at the bulk, respectively, of the CNT [151], see Fig. 3-19.

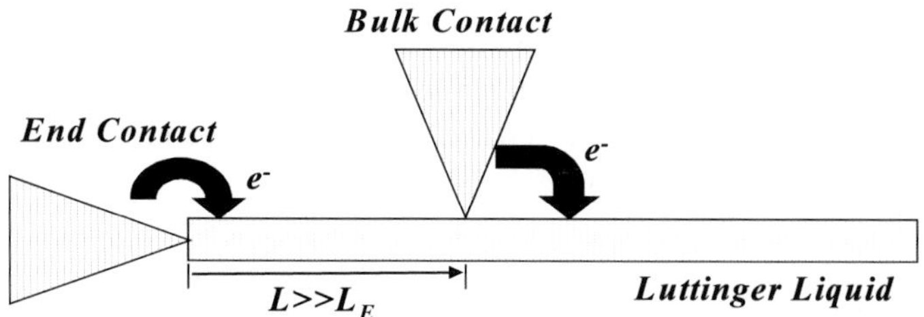

Figure 3-19. 3D-1D contact to carbon nanotube. (*After* [151].)

Similarly, efforts have been expended, and are being vigorously pursued, to uncover the predicted spin-charge separation. These include proposals to directly excite Lüttinger liquid behavior in CNTs by impressing microwave voltage waves in CNTs acting as transmission lines [149].

3.2.1.2 Superconductors

The phenomenon of superconductivity manifests itself as the drop in the electrical resistance of metals and alloys at sufficiently low temperatures, accompanied by the inhibition of magnetic fields from penetrating inside of them [28]. Conversely, a material in the superconducting state loses this property when its temperature is raised past a critical temperature, T_c, or it is exposed to a high critical magnetic field H_c [28]. We discuss the principles of superconductivity here, mainly because of the importance of superconductors (materials exhibiting superconductivity) as an alternative means of implementing quantum bits (*qubits*). Our point of departure in discussing superconductivity is the concept of superfluidity, from which it may be understood in an intuitive fashion.

3.2.1.2.1 Superfluidity

Superfluidity refers to the property exhibited by a superfluid, i.e., a liquid that flows without friction. A successful explanation of superfluidity was put forth by Landau [153], [154]. Landau's reasoning was as follows [131]. If one assumes that the Bose quantum fluid of mass M is in its ground state at absolute zero, and flowing within a capillary tube with velocity v, and energy $\frac{1}{2}\mathrm{Mv}^2$, then, in a coordinate system anchored in the fluid, the fluid would be at rest and the capillary would appear to be moving at a velocity –v. If friction emerges between the capillary and the fluid, then the part of the latter in contact with the tube would no longer be at rest, but would begin to be carried along by the capillary wall. However, since this part of the fluid would no longer be at rest, the act of it being carried along by the tube wall must induce excitations from its ground state. These excitations, in turn, would manifest as changes in its energy and momentum, E and p, so that the fluid's total energy would now be $\mathrm{E}+\vec{\mathrm{p}}\cdot\vec{\mathrm{v}}+\frac{1}{2}\mathrm{Mv}^2$. Upon excitation, the fluid itself would lose energy. Therefore, energy change must be negative, i.e.,

$$\mathrm{E}+\vec{\mathrm{p}}\cdot\vec{\mathrm{v}}<0. \tag{88}$$

Since the fluid is a quantum system of Bose particles, its energy is quantized and must change discretely. The smallest energy excitation, therefore, is that for which $\mathrm{E}+\vec{\mathrm{p}}\cdot\vec{\mathrm{v}}$ is a minimum, which occurs when $\vec{\mathrm{p}}$ and $\vec{\mathrm{v}}$ are opposite. This means that one must have,

$$\mathrm{E}-\mathrm{pv}<0 \qquad \text{or} \qquad \mathrm{v}>\frac{\mathrm{E}}{\mathrm{p}}. \tag{89}$$

This equation sets the minimum velocity at which excitations would begin, as the *critical velocity*,

$$v_c > min\left(\frac{E}{p}\right). \quad (90)$$

In particular, if $v_c \neq 0$, then it is possible for the fluid to flow free of excitations, i.e., without friction/dissipation, as long as $v < v_c$. This is the so-called Landau's criterion for superfluidity. This condition is maintained as long as v is less than the speed of sound. This insight, led Landau to propose that the low energy excitations of the superfluid ground state should consist of two types of particles, namely, phonons and "rotons." Phonons being quantized sound waves, with an energy dispersion,

$$E = Sp, \quad (91)$$

where S is the speed of sound and p the momentum, and rotons being quantized rotational motion (vortices), with an energy dispersion,

$$E = \Delta + (p - p_0)^2/2m_{\text{eff}}. \quad (92)$$

At temperatures above absolute zero, the fluid will be excited by thermal energy. Therefore, it will be possible for some of the thermally excited fluid particles to achieve velocities greater than v_c and will, consequently, experience friction. Under these circumstances, the fluid will be composed of these normal particles and superfluid particles, resulting in a mass current given by,

$$\vec{j} = \rho_n \vec{v}_n + \rho_s \vec{v}_s, \quad (93)$$

where ρ_n and v_n are the mass density and velocity of the normal fluid, and ρ_s and v_s those of the superfluid. If one assumes that the whole fluid flows with velocity $\vec{v}_n = \vec{v}_s = \vec{v}$, then the total mass current may be written as,

$$\vec{j} = (\rho_n + \rho_s)\vec{v} = \rho\vec{v}. \quad (94)$$

One of the fundamental properties of a superfluid derives from the fact that, since it possesses no excited particles, its momentum doesn't change and, consequently, it can't exert a force on a body immersed in it. Flow with this property, denoted "potential flow," is mathematically characterized by the equation,

$$\nabla \times \mathrm{v}_s = 0 . \tag{95}$$

Eq.(95) signifies that a superfluid is irrotational, i.e., it exhibits no vorticity. The quintessential example of a superfluid is embodied by a Bose liquid, which consists of atoms of integral-value spins, in particular, liquid helium (He^4), which does not solidify at absolute zero and flows through capillaries without dissipation.

Landau's arguments, presented above, while successfully explaining liquid helium behavior, were of an intuitive and phenomenological nature. Elements for a first-principles theory to explain superfluid behavior began taking shape with observations by Fritz London [155], to the effect that the constitution of He atoms, which are composed of an even number of elementary particles (2 protons, 2 neutrons, and 2 electrons) suggested that they should be described by a symmetric wavefunction and, consequently, should obey Bose statistics, together with the earlier observation by Einstein that, at appropriately low temperatures and mass and density conditions, a gas of non-interacting Bose particles condenses with the remarkable property that a nonzero fraction of the condensed atoms occupies a single one-particle state. Such a state, in particular, is a coherent state and has come to be known as a Bose-Einstein condensate (BEC) [155]. A fundamental theory capturing this behavior is the Gross-Pitaevskii (GP) model. The GP equation models the general Bose gas by the equation [78],

$$i\hbar \frac{\partial \psi}{\partial \mathrm{t}} = -\frac{\hbar^2}{2m} \nabla^2 \psi + U_{mf} \psi , \tag{96}$$

where m is particle mass,

$$U_{mf} = e^2 \int \frac{|\psi(x)|^2 \, dx'}{|x - x'|} , \tag{97}$$

is the mean field for Coulomb interaction between atoms, and may be expressed as,

$$U_{mf} = \int \mathrm{V}(x - x')|\psi(x')|^2 \, dx' = \int g\delta(x - x')|\psi(x')|^2 \, dx' = g|\psi(x')|^2 . \tag{98}$$

Substituting (98) into (96) one obtains a nonlinear Schrödinger equation,

$$i\hbar \frac{\partial \psi}{\partial \mathrm{t}} = -\frac{\hbar^2}{2m} \nabla^2 \psi + g|\psi(x)|^2 \psi(\mathrm{x}) = -\frac{\hbar^2}{2m} \nabla^2 \psi + g\psi^* \psi^2 . \tag{99}$$

Condensation is captured when in (99) one imposes the conditions for obtaining the lowest possible state, ψ_0, namely, that the wave function be homogeneous, i.e.,

$$\nabla^2\psi_0 \to 0. \tag{100}$$

This leads to the relation,

$$\left|\psi_0\right|^2 = n = \frac{N}{V}, \tag{101}$$

where N is the number of atoms and V is the volume. In turn, substitution of (100) and (101) into (99) leads to a simplified equation of motion, namely,

$$i\hbar\frac{\partial\psi_0}{\partial t} = gn\psi_0, \tag{102}$$

with a solution of the form,

$$\psi_0 = Ce^{-\frac{gnt}{\hbar}}. \tag{103}$$

The dispersion relation for low-level excitations are obtained by linearizing (99), in particular, writing $\psi = \psi_0 + \chi$, where $\chi << \psi_0$, and substituting into (99), one obtains,

$$i\hbar\frac{\partial\chi}{\partial t} = -\frac{\hbar^2}{2m}\nabla^2\chi + 2g\psi_0{}^2\chi + g\psi_0{}^2\chi^*. \tag{104}$$

Since this equation contains the two unknowns χ and χ^*, we generate a second equation by taking its complex conjugate,

$$-i\hbar\frac{\partial\chi^*}{\partial t} = -\frac{\hbar^2}{2m}\nabla^2\chi^* + 2g\psi_0{}^2\chi^* + g\psi_0{}^2\chi. \tag{105}$$

Then, postulating solutions of the form,

$$\chi \sim \xi e^{-iEt+ipx}, \tag{106}$$

and

$$\chi^{*} \sim \eta e^{-iEt+ipx}, \tag{107}$$

substituting them into (104) and (105), together with (101), and solving the resulting system of equations for E, one obtains the result,

$$E(p)=\sqrt{S^{2}p^{2}+\frac{p^{4}}{4m^{2}}}\,. \tag{108}$$

This is the dispersion relation of a superfluid. Expressing the fluid velocity in terms of it, we obtain,

$$\mathrm{v}=\frac{E}{p}=\sqrt{S^{2}+\frac{p^{2}}{4m^{2}}}\,. \tag{109}$$

This equation has a positive minimum, occurring at $p \to 0$, and given by the constant velocity *S*. Since this velocity is independent of momentum, *E(p)* must contain an energy gap. An energy gap in its spectrum, thus, is another manifestation of superfluidic behavior.

The zero-vorticity property of a superfluid is derived from first principles as follows. From (103) it may be seen that the wave function for the Bose condensate in its lowest energy state is a one-particle complex wave. Generalizing this expression to,

$$\psi(x)=|\psi|e^{i\chi(x)}, \tag{110}$$

one can express the mass density as $\rho = m|\psi|^{2}$, where $\psi(x)$ and the current are related, as usual, by,

$$\vec{\mathrm{j}}=-\frac{i\hbar}{2}\left(\psi^{*}\nabla\psi-\psi\nabla\psi^{*}\right). \tag{111}$$

It then follows that, inserting (110) into (101) one obtains,

$$\vec{\mathrm{j}}=\hbar|\psi|^{2}\nabla\chi=\rho\frac{\hbar}{m}\nabla\chi, \tag{112}$$

which, upon comparison with (94) yields,

$$\vec{v} = \frac{\hbar}{m}\nabla\chi, \tag{113}$$

that is, the velocity is related to the phase, χ, of the wave function, so one can rewrite (113) as,

$$\vec{v} = \nabla\phi, \tag{114}$$

which clearly expresses that the flow is a potential flow, since the curl of any gradient is zero, and the potential is given by,

$$\phi = \frac{\hbar}{m}\chi. \tag{115}$$

A further phenomenon accomplanying superfluidity, and elucidated by first-principles considerations, pertains to the dynamics of superfluids when placed in a rotating container. In particular, it is experimentally found, Fig. 3-19, in a vessel containing a mixture of normal and superfluid components, and rotating at an angular velocity $\vec{\Omega}$, that the dynamic behavior of the two components is quite different. On the one hand, as is expected from classical hydrodynamics, the normal component rotates with the vessel (i.e., it is carried along with the vessel due to friction), so that it acquires an eddy current $\vec{v}_n = \vec{\Omega}\times\vec{r}$, and this velocity, in turn, gives rise to an accompanying vortex, since $\vec{\nabla}\times\vec{v}_n = 2\vec{\Omega}$, see Fig. 3-19. The superfluid component, on the other hand, becomes populated by a distribution of vortices. This appearance of vortices in the superfluid component would appear to contradict the fundamental condition for superfluidity of zero vorticity, see Eq.(95). The clue to this behavior was to be found in the recognition that potential flow, characterized by (95), may also be obtained whenever the equivalent form, based on Stokes' theorem,

$$\oint\vec{v}_s d\vec{r} = 0, \tag{116}$$

is satisfied. In particular, if the potential of the rotating fluid is proportional to the angle, see Fig. 3-20, so that one has,

$$\phi = \frac{\Gamma}{2\pi}\alpha, \tag{117}$$

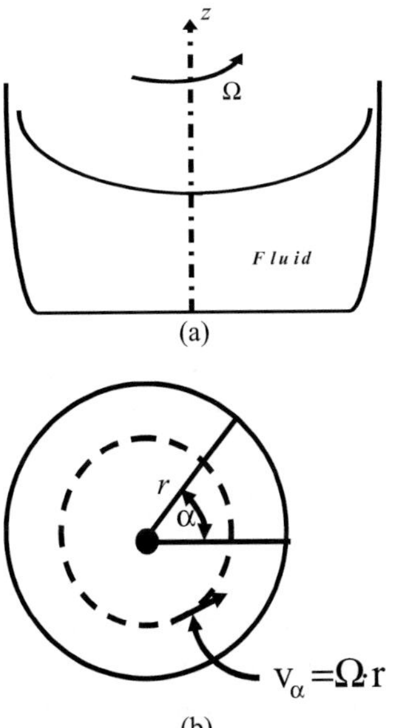

Figure 3-20. (a) Normal fluid in rotating vessel acquires meniscus with shape depending only on angular velocity Ω. Top view of fluid-containing vessel rotating with angular velocity Ω. The normal fluid acquires an eddy current with velocity v_α.

then the fluid velocity may be calculated as,

$$v_\alpha(r) = \frac{1}{r}\frac{\partial}{\partial\alpha}\phi = \frac{\Gamma}{2\pi r}, \tag{118}$$

and, since the velocity decays with distance, this is the profile of a vortex. Now, calculation of the circulation of this vortex gives,

$$\oint v_\alpha \cdot dl = \Delta\phi = \Gamma. \tag{119}$$

Examination of Eq. (119) reveals that if the circulation (potential change) is zero, one still has the conflict between the mathematical violation of vorticity and the experimental observation of vortices. However, if the angle α is not uniquely defined, except up to modulus 2π, then it would be possible to reconcile the two if the potential ϕ were not single-valued. This, in turn, would be the case if the phase of the wavefunction was not unique, but also defined modulo 2π, so that $\Delta\chi = 2\pi N$. In this case, the circulation (119) would be expressed as,

$$\Gamma = \Delta\phi = \frac{2\pi\hbar}{m} N , \tag{120}$$

that is, it would be *quantized*. Thus, a change in potential of $2\pi\hbar/m$ would bring it to the point of departure, due to its non-single-valuedness, yet would allow a non-zero vorticity due to its finiteness. The quantum nature of a superfluid contained in a rotating vessel manifests, therefore, in that its circulation becomes quantized. One remarkable aspect of a rotating vessel containing a superfluid pertains to the shape of its meniscus. In particular, from the fact that a normal fluid in a vessel of area A rotating at an angular velocity Ω has a circulation $2\Omega\mathrm{A}$, and that a superfluid on the same vessel would have a circulation $\Gamma\nu\mathrm{A}$, where ν is the density of vortices per unit area, one finds, equating circulations, that the $\Omega = \Gamma\nu/2$. This signifies, that although the superfluid would not necessarily be rotating, due to the appearance of vortices, the shape of its meniscus will be the same as that of a normal fluid rotating at an angular velocity Ω. In other words, one can *simulate* the effect of rotation on a normal fluid by a population of vortices.

The fact that the circulation of a superfluid contained in a rotating vessel is quantized means that the vessel must reach a certain minimum angular velocity, the *critical angular velocity*, Ω_c, and rotational energy before the vortices begin to be created. From the ratio of vortex energy to vortex angular momentum it can be shown that,

$$\Omega_c = \frac{\hbar}{mR^2} , \tag{121}$$

where R is the vessel radius. Figure 3-21 shows a picture of vortices in a superfluid.

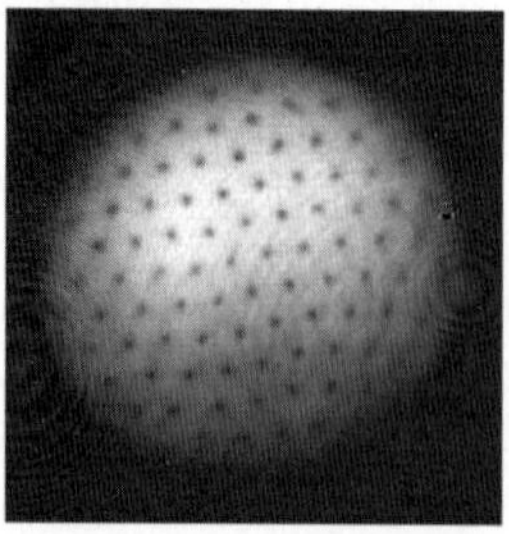

Figure 3-21. Observation of vortex lattices. The examples shown contains approximately 80, vortices. The vortices have "crystallized" in a triangular pattern. Reprinted with permission from [156]. Copyright 2001 AAAS.

3.2.1.2.2 Superconductivity

Our understanding of superfluidity, gained in the previous section, facilitates that of superconductivity. Superconductivity, the absence of electrical resistance to electron transport, may be conceptually visualized as the "superfluidity of electrons". A qualitative analogy between these two phenomena may be summarized as follows. Whereas a superfluid embodies a boson condensate of, e.g., helium atoms, a superconductor, on the other hand, embodies boson condensates of, e.g., bound electron pairs. Electrons, as is known, due to the Coulomb force of repulsion between them, do not, strictly speaking, condense. However, under certain circumstances, an effective binding force may be present that overcomes the force of repulsion between electron pairs and turns these pairs, effectively, into bosons. These electron pairs, which behave as bosons, are called *Cooper pairs* and have zero spin (just as the helium atoms). Thus, while a boson condensate of helium atoms may behave as a superfluid, under appropriate circumstances, and when it does so it exhibits transport without friction, so too a condensate of an aggregate of Cooper pairs, behaves as a superconductor. Continuing with the analogy, while superfluid transport exists for velocities less than a critical velocity, $v_c \sim min(E/p)$, so too superconductive transport exists below a critical velocity $v_c \sim (\Delta/p_0)$, where 2Δ in this case is the binding energy of a Cooper pair. Finally, while dissipation and fluid vortices (rotons) appear above v_c in the superfluid, so too ohmic dissipation and so-called vortex states, i.e., circulation of superconducting currents in vortices throughout the system, appear beyond v_c in the superconductor. With these qualitative preliminaries, we next address the salient aspects of superconductivity, namely, the criterion for superconductivity in light of its conceptual relationship to superfluidity, the binding energy of Cooper pairs, the inhibition of a magnetic field inside superconducting materials, the conditions for the extinction of superconductivity.

In analogy with (105), the equation for a single electron moving in a superconductor may be written as,

$$i\hbar \frac{\partial \psi_\sigma(x,t)}{\partial t} = -\frac{\hbar^2}{2m}\nabla^2 \psi_\sigma + g\psi^*_{\bar{\sigma}}\psi_{\bar{\sigma}}\psi_\sigma, \tag{122}$$

where g represents charge, $\sigma = \uparrow$ or $\downarrow$ represents the spin state, and $\psi^*_{\bar{\sigma}}\psi_{\bar{\sigma}}$ is a 2-index summation that embodies the density from all spins. In this context, the wave function of a pair of electrons is a product given by,

$$\Psi_{\bar{\sigma}\sigma}(x_1,x_2)=\psi_{\bar{\sigma}}(x_1)\psi_{\sigma}(x_2). \tag{123}$$

Being the wave function of a boson, $\Psi_{\bar{\sigma}\sigma}(x_1,x_2)$ must satisfy Pauli's exclusion principle, wheraby it must be anti-symmetric. Furthermore, since spins and spatial coordinates operate in different (tensor) spaces, the wave function must be a product of a spin-dependent factor, and a coordinate-dependent factor, i.e.,

$$\Psi_{\bar{\sigma}\sigma}(x_1,x_2)=E_{\bar{\sigma}\sigma}f(x_1,x_2)=E_{\bar{\sigma}\sigma}\cdot\frac{\Delta}{g}, \tag{124}$$

where $E_{\bar{\sigma}\sigma}$ is the anti-symmetric spin-dependent factor. With this definition, one can rewrite (124) as,

$$i\hbar\frac{\partial\psi_{\sigma}(x,t)}{\partial t}=-\frac{\hbar^2}{2m}\nabla^2\psi_{\sigma}+\Delta E_{\bar{\sigma}\sigma}\psi_{\bar{\sigma}}^{*}. \tag{125}$$

Following the same procedure as in the previous section, the dispersion relation is obtained from the set of equations,

$$i\hbar\frac{\partial\psi_{\sigma}(x,t)}{\partial t}=-\frac{\hbar^2}{2m}\nabla^2\psi_{\sigma}-E_F\psi_{\sigma}+\Delta E_{\bar{\sigma}\sigma}\psi_{\bar{\sigma}}^{*}, \tag{126a}$$

and

$$-i\hbar\frac{\partial\psi_{\sigma}^{*}(x,t)}{\partial t}=-\frac{\hbar^2}{2m}\nabla^2\psi_{\bar{\sigma}}^{*}-E_F\psi_{\bar{\sigma}}^{*}+\Delta^{*}E_{\bar{\sigma}\sigma}\psi_{\bar{\sigma}}, \tag{126b}$$

where the energy is now referred to the Fermi energy. Then, postulating solutions of the form,

$$\psi_{\bar{\sigma}}\sim\eta_{\sigma}e^{-iEt/\hbar+ipx/\hbar}, \tag{127a}$$

and

$$\psi_{\bar{\sigma}}^{*}\sim\zeta_{\bar{\sigma}}e^{-iEt/\hbar+ipx/\hbar}, \tag{127b}$$

it can be shown, upon substitution on (126), that the set of equations,

$$E\eta_{\sigma} = \left(\frac{p^{2}}{2m} - E_{\mathrm{F}}\right)\eta_{\sigma} + \Delta E_{\bar{\sigma}\sigma}\zeta_{\bar{\sigma}}, \tag{128a}$$

and

$$-E\zeta_{\bar{\sigma}} = \left(\frac{p^{2}}{2m} - E_{\mathrm{F}}\right)\zeta_{\bar{\sigma}} + \Delta^{*} E_{\sigma\bar{\sigma}}\eta_{\sigma}, \tag{128b}$$

is obtained. Solving (128) for E one obtains $E = \pm\sqrt{|\Delta|^{2} + \mathrm{v}_{\mathrm{F}}^{2}(p - p_{F})^{2}}$.

This is the dispersion relation for superconducting electrons. It represents a parabola with a minimum at $p = p_{F}$, corresponding energy Δ, and energy gap 2Δ. Therefore, application of the Landau criterion for superfluidity, to the present case of superconductivity, yields the critical velocity, $\mathrm{v}_{\mathrm{c}} = (\Delta / p_{F})$, below which electron transport experiences no electrical resistance, i.e., is superconductive. Next, we address the formation of Cooper pairs.

In exploiting the superfluid physics analogy to describe superconductivity, one must confront the issue of explaining how electrons, which would ordinarily be precluded from binding, due to Coulomb's repulsion force, would bond/condense to form bosons. The clue to this possibility was advanced by the discovery that [157], [158] in superconducting elements, the product of the square root of their isotopic mass and the critical temperature, $M^{1/2}T_{c}$, is a constant. This experimental fact, in turn, was interpreted by Fröhlich [154] to mean that the properties of the zero-point or thermal lattice phonons, were involved in superconductivity and, in particular, that electrons residing within the crystal lattice were capable, via interactions mediated by these phonons, of attracting one another. This phenomenon is demonstrated next.

To determine the nature of the phonon-mediated electron-electron interaction, we assume the coexistence of phonons and electrons is described by a Hamiltonian consisting of three terms, namely, the energy of the electrons, the energy of the phonons, and the energy of interaction between electrons and phonons, respectively. The first two terms are captured by the "unperturbed" Hamiltonian:

$$H_{0} = \sum_{\vec{k},\sigma} E_{\vec{k},\sigma} c_{\vec{k},\sigma}^{+} c_{\vec{k},\sigma} + \sum_{\vec{q}} \hbar\omega_{\vec{q}} a_{\vec{q}}^{+} a_{\vec{q}} . \tag{129}$$

The third term is the familiar electron-phonon interaction [159], in which an acoustic phonon distorts the lattice and, as a consequence, produces a grating in the band edges which, in turn, causes electrons to scatter off of it. This interaction is captured by the interaction potential for acoustic phonons given by,

$$U_{AP}(\vec{r},t) = D\vec{\nabla}u(\vec{r},t), \tag{130}$$

where D is the deformation potential and,

$$u(\vec{r},t) = \sqrt{\frac{\hbar}{2\rho V\omega(\vec{q})}}\left(a_{\vec{q}}e^{i[\vec{q}\cdot\vec{r}-\omega(\vec{q})t]} + a_{\vec{q}}^{*}e^{-i[\vec{q}\cdot\vec{r}-\omega(\vec{q})t]}\right), \tag{131}$$

the lattice displacement. The pertinent energy of interaction is,

$$H_{ep} = \int d\vec{r}\,\Psi^{+}(\vec{r})U_{AP}(\vec{r})\Psi(\vec{r}), \tag{132}$$

where,

$$\Psi(\vec{r}) = \sum_{\vec{k}} c_{\vec{k}} e^{i\vec{k}\cdot\vec{r}}\phi_{\vec{k}}(\vec{r}), \tag{133}$$

is the unperturbed one-electron Block state. With these definitions, the first-order electron-phonon interaction may be written as,

$$H' = iD\sum_{\vec{k},\sigma\vec{q}} c_{\vec{k}+\vec{q},\sigma}^{+} c_{\vec{k},\sigma}\left(a_{\vec{q}} - a_{\vec{q}}^{*}\right). \tag{134}$$

The Hamiltonian describing the electron-phonon system, then, is given by,

$$H = \sum_{\vec{k},\sigma} E_{\vec{k},\sigma} c_{\vec{k},\sigma}^{+} c_{\vec{k},\sigma} + \sum_{\vec{q}} \hbar\omega_{\vec{q}} a_{\vec{q}}^{+} a_{\vec{q}} + iD\sum_{\vec{k}\vec{q}} c_{\vec{k}+\vec{q}}^{+} c_{\vec{k}}\left(a_{\vec{q}} - a_{\vec{q}}^{*}\right). \tag{135}$$

Now, to determine the nature of the electron-electron interaction, we have to transform (135) into a Hamiltonian that *does not* contain the *O(D)* term, i.e., in which the phonon coordinates are eliminated and only electron-electron interaction terms are present. This is accomplished by transforming (135) into a new Hamiltonian given by $\widetilde{H} = e^{-S}He^{S}$, and so choosing *S* that $\widetilde{H}$

contains no off-diagonal terms of *O(D)*. In particular, if we take $H = H_0 + H'$, where H_0 gives rise to the solutions $|n\rangle$ when $H' = 0$, so that $H_0|n\rangle = E_n|n\rangle$, then $\widetilde{H}$ may be expanded as, [131],

$$\begin{aligned}\widetilde{H} &= e^{-S}(H_0 + H')e^{S} = \left(1 - S + \frac{S^2}{2} + ...\right)(H_0 + H')\left(1 + S + \frac{S^2}{2} + ...\right) \\ &= H + [H,S] + \frac{1}{2}[[H,S],S] + ... \\ &= H_0 + H' + [H_0,S] + \frac{1}{2}[H' + [H_0,S],S] + \frac{1}{2}[H',S] + ...\end{aligned} \quad (136)$$

If we select $H' + [H_0,S] = 0$, then the second and third terms in (136) vanish and we have a prescription for *S*, namely,

$$\langle n'|H'|n\rangle + (E_{n'} - E_n)\langle n'|S|n\rangle = 0 \Rightarrow \langle n'|S|n\rangle = \frac{\langle n'|H'|n\rangle}{E_{n'} - E_n}, \quad (137)$$

which yields the desired Hamiltonian as,

$$\widetilde{H} = H_0 + \frac{1}{2}[H',S] + O(S^2). \quad (138)$$

Now, in this diagonal formulation, effective electron-electron interaction is elucidated by considering the case in which the perturbation H' causes the following transitions: Either the electron in state k emits a phonon –q and this is absorbed by the electron in state k', or the electron in state k' emits a phonon q and this is absorbed by the electron in state k. These transitions may be mathematically represented as occurring from an initial state $|i\rangle$ to a final state $|f\rangle$ via a virtual state $|m\rangle$, in terms of which the expectation value of the commutator in (138) may be expressed as,

$$\langle f|[S,H']|i\rangle = \sum_m \left(\langle f|S|m\rangle\langle m|H'|i\rangle - \langle f|H'|m\rangle\langle m|S|i\rangle\right). \quad (139)$$

Following [154], consideration of the phonon system at absolute zero, so that one of the phonon states refers to the vacuum, the matrix element calculation (134) over the phonon operators yields, without loss of generality,

$$\left\langle 1_q \middle| S \middle| 0 \right\rangle = \sum_{\vec{k}} \frac{i c^+_{\vec{k}-\vec{q}} c_{\vec{k}} D}{E_{\vec{k}} - E_{\vec{k}-\vec{q}} - \hbar\omega_{\vec{q}}}, \tag{140a}$$

and

$$\left\langle 0 \middle| S \middle| 1_{\vec{q}} \right\rangle = \sum_{\vec{k}'} \frac{i c^+_{\vec{k}'+\vec{q}} c_{\vec{k}'} D}{E_{\vec{k}'} + \hbar\omega_{\vec{q}} - E_{\vec{k}'-\vec{q}}}. \tag{140b}$$

Substituting (140) into (139) one obtains,

$$\left\langle f \middle| \left[S, H'\right] \middle| i \right\rangle = \frac{1}{2} D^2 \sum_{\vec{q}\vec{k}\vec{k}'} c^+_{\vec{k}'+\vec{q}} c_{\vec{k}'} c^+_{\vec{k}'+\vec{q}} c_{\vec{k}'} \left(\frac{1}{E_{\vec{k}} - E_{\vec{k}-\vec{q}} - \hbar\omega_{\vec{q}}} - \frac{1}{E_{\vec{k}'} + \hbar\omega_{\vec{q}} - E_{\vec{k}'-\vec{q}}} \right). \tag{141}$$

Realizing that, due to energy conservation, $E_{\vec{k}'} - E_{\vec{k}'-\vec{q}} = E_{\vec{k}+\vec{q}} - E_{\vec{k}}$, (142) may be simplified to yield,

$$H'' = D^2 \sum_{\vec{q}} \sum_{\vec{k}\vec{k}'} \frac{\hbar\omega_{\vec{q}} c^+_{\vec{k}'+\vec{q}} c_{\vec{k}'} c^+_{\vec{k}-\vec{q}} c_{\vec{k}'}}{\left(E_{\vec{k}} - E_{\vec{k}-\vec{q}}\right)^2 - \omega^2_{\vec{q}}}. \tag{143}$$

Equation (140) reveals that in circumstances when $\left(E_{\vec{k}} - E_{\vec{k}-\vec{q}}\right)^2 < \omega^2_{\vec{q}}$, this term is negative, thus embodying an electron-electron interaction that is attractive, and that gives rise to the bosonic behavior mentioned previously.

Having shown that it is physically possible for a pair of electrons to attract one another in the presence of a phonon, the next question before us is to determine the binding energy of the pair. As usual, this is obtained from the energy eigenvalues of Schrödinger equation, $H\psi = E\psi$. Towards this end, we begin by expressing the Hamiltonian,

$$H = \frac{\vec{p}_1^{\,2}}{2m} + \frac{\vec{p}_2^{\,2}}{2m} + V\left(\left|\vec{r}_1 - \vec{r}_2\right|\right), \tag{144}$$

where the potential $V\left(\left|\vec{r}_1 - \vec{r}_2\right|\right)$ models the interaction (143), in the center-of-mass and relative-motion coordinates, i.e.,

$$H = \frac{\vec{P}^2}{4m} + \frac{\vec{p}^{\,2}}{2m} + V\left(\left|\vec{r}\right|\right), \tag{145}$$

with $\vec{R} = (\vec{r}_1 + \vec{r}_2)/2$, $\vec{r} = \vec{r}_1 - \vec{r}_2$, $\vec{P} = \vec{p}_1 + \vec{p}_2$, and $\vec{p} = (\vec{p}_1 - \vec{p}_2)/2$. Then, expressing the solution as,

$$\psi = e^{i\vec{K}\cdot\vec{R}} \sum_{\vec{k}} h_k e^{i\vec{k}\cdot\vec{r}}, \tag{146}$$

and taking into consideration the symmetry properties of the problem, in particular, upon interchange of $\vec{r}_1$ and $\vec{r}_2$, $\vec{R} \to \vec{R}$, $\vec{r} \to -\vec{r}$, and $h_{-\vec{k}} = h_{\vec{k}}$, and in the frame of reference in which the system is at rest $\vec{K} = 0$, substitution of (146) into the Schrödinger equation, $H\psi = E\psi$, yields,

$$\begin{gathered}\sum_{\vec{k}} h_{\vec{k}} \frac{\hbar^2 k^2}{2m} e^{i\vec{k}\cdot\vec{r}} + \sum_{\vec{k}'} V(|\vec{r}|) e^{i\vec{k}'\cdot\vec{r}} h_{\vec{k}'} = E \sum_{\vec{k}} h_{\vec{k}} e^{i\vec{k}\cdot\vec{r}} \\ \Rightarrow \left(E - \frac{\hbar^2 k^2}{2m}\right) h_{\vec{k}} = \sum_{\vec{k}'} \frac{1}{\Omega} \int_{(\Omega)} d\vec{r}\, e^{-i\vec{k}\cdot\vec{r}} V(|\vec{r}|) e^{i\vec{k}'\cdot\vec{r}} h_{\vec{k}'} = \sum_{\vec{k}'} V_{\vec{k}\vec{k}'} h_{\vec{k}'}\end{gathered} \tag{147}$$

Since the electron-electron interaction is mediated by phonons, and the phonon energies lie between 0 and $\hbar\omega_D$, where ω_D is the Debye energy, the electrons will be under the influence of the binding potential as long as the their excitation energy of the pair is lower than the Debye energy, i.e., $\left|\varepsilon_{\vec{k}} - \varepsilon_{\vec{k}'}\right| < \hbar\omega_D$, $\varepsilon_k = \hbar^2 k^2 / 2m$. In this context, we have,

$$V_{\vec{k}\vec{k}'} = -V \tag{148}$$

and we can write (147) as,

$$\left(E - \frac{\hbar^2 k^2}{2m}\right) h_{\vec{k}} = -V \sum\nolimits'_{\vec{k}'} h_{\vec{k}'}, \tag{149}$$

which, may be expressed as,

$$\sum\nolimits'_{\vec{k}} h_{\vec{k}} = \sum\nolimits'_{\vec{k}'} h_{\vec{k}'} \sum\nolimits'_{\vec{k}'} \frac{V}{\left(E - \dfrac{\hbar^2 k^2}{2m}\right)}, \tag{150}$$

which may be factored as,

$$\Rightarrow \left(\sum_{\vec{k}}' h_{\vec{k}}\right)\left(1+\sum_{\vec{k}'}' \frac{V}{\left(E-\frac{\hbar^2 k^2}{2m}\right)}\right) = 0\,, \tag{151}$$

from where we get,

$$\Rightarrow \sum_{\vec{k}'}' \frac{V}{\left(E-\frac{\hbar^2 k^2}{2m}\right)} = -1. \tag{152}$$

Replacing summation by integration we obtain,

$$V\int_{E_F}^{E_F+\hbar\omega_D} d\varepsilon N(\varepsilon)\frac{1}{\varepsilon - E} = VN(0)\int_{E_F}^{E_F+\hbar\omega_D} \frac{d\varepsilon}{\varepsilon - E} = 1\,, \tag{153}$$

where N(0) is the density of electronic states for a single spin population in the normal metal [64]. Upon carrying out the integration we get,

$$VN(0)\ln\frac{E_F+\hbar\omega_D - E}{E_F - E} = 1\,, \tag{154}$$

which may be solved by the energy of the pair,

$$E_F - E = \frac{\hbar\omega_D}{e^{\frac{1}{VN(0)}} - 1}. \tag{155}$$

Clearly, (155) denotes a system energy that is below the Fermi energy, therefore, we have a bound state. Observing that the reduced mass m and the electron mass m_0 are related by $m = m_0/2$, effecting the corresponding substitution $\hbar^2 k^2/2m \rightarrow \hbar^2 k^2/m_0 = 2\varepsilon_k$, and repeating the operations of (153)-(154) one obtains the result,

$$E = 2E_F - \frac{2\hbar\omega_D}{e^{\frac{2}{VN(0)}} - 1}. \tag{156}$$

The zero-temperature binding energy (gap) is given by,

$$2\Delta = \frac{2\hbar\omega_D}{e^{\frac{2}{VN(0)}} - 1} = E_b. \tag{157}$$

The binding energy (157) determines how far apart the electrons forming a Cooper pair may separate while still acting as bound. In this context, the radius R of a Cooper pair has been estimated as [160],

$$R \sim \frac{\hbar^2 k_F}{m E_b}, \tag{158}$$

which, numerically, is of the order of $1\mu m$. The implication of the binding energy is as follows. At absolute zero, an energy greater than the binding energy is required to separate Cooper pairs and, thus, create excited electrons which are generated in pairs. At energies close to this threshold, E_b, the current will consist of both Cooper pairs and single (normal) electrons resulting from the breaking of the pairs, giving rise to a two-fluid model transport. Abrikosov has shown that as the temperature increases E_b decreases until it reaches zero a the critical temperature, T_c. This is temperature dependence is given by,

$$E_b = 3.06\sqrt{T_c(T_c - T)}. \tag{159}$$

Next, we consider the phenomenon of magnetic field exclusion from a superconductor. We examine the supercurrent in a superconductor containing a density of n_s electrons moving with velocity v_s and, thus, given by $J_s = en_s v_s$, in the presence of a vector potential field $\vec{A}'$. In general, the particle velocity in a vector potential is given by,

$$\vec{v} = \frac{1}{M}\left(\vec{p} - \frac{q}{c}\vec{A}'\right). \tag{160}$$

In the case of the superconductor, $M = 2m_e$, and $q = 2e$. If we let Ψ be the wavefunction of the electron pair (boson), then we can express (160) as,

$$
\begin{aligned}
|\Psi|^2 \vec{\mathrm{v}} &= \frac{1}{\mathrm{M}}\left(-\mathrm{i}\hbar\Psi^*\nabla\Psi - \frac{q}{c}|\Psi|^2 \vec{A}'\right) \\
&= \frac{1}{2m_e}\left(-\mathrm{i}\hbar\Psi^*\nabla\Psi - \frac{2e}{\mathrm{c}}|\Psi|^2 \vec{A}'\right)
\end{aligned} \tag{161}
$$

Now, writing the complex wave function as $\Psi = |\Psi| e^{i\chi}$, where χ is a space-dependent phase, and substituting into (161) we obtain,

$$
\vec{v}_s = \frac{\hbar}{2m_e}\nabla\chi - \frac{e}{m_e c}\vec{A}'. \tag{162}
$$

This equation reveals that, even if $\nabla\chi = 0$, current flow may be excited by the vector potential. In fact, since $\vec{\mathrm{B}} = \nabla\times\vec{\mathrm{A}}$, we may redefine $\vec{\mathrm{A}}'$ to include the phase, without changing $\vec{\mathrm{B}}$, i.e.,

$$
\vec{\mathrm{A}} = \vec{\mathrm{A}}' + \frac{\hbar \mathrm{c}}{2\mathrm{e}}\nabla\chi, \tag{163}
$$

from where we get,

$$
\vec{v}_s = -\frac{e}{m_e c}\vec{A}'. \tag{164}
$$

The supercurent, then, is given by,

$$
\vec{J}_s = -\frac{e^2 n_s}{m_e c}\vec{A}'. \tag{165}
$$

The effects of a superconductor on a magnetic field inside its bulk follow from from substituting (164) into the equation (165),

$$
\nabla\times\vec{\mathrm{B}} = \frac{4\pi}{\mathrm{c}}J_s, \tag{166}
$$

and taking its curl, i.e.,

$$
\nabla\times\nabla\times\vec{\mathrm{B}} = \frac{4\pi}{\mathrm{c}}\nabla\times J_s = -\frac{4\pi \mathrm{e}^2 \mathrm{n_s}}{\mathrm{m_e c^2}}\vec{\mathrm{B}}. \tag{167}
$$

Since $\nabla \cdot \vec{B} \equiv 0$, (167) becomes,

$$\nabla^2 \vec{B} - \frac{4\pi e^2 n_s}{m_e c^2} \vec{B} = 0, \tag{168}$$

which may be rewritten as,

$$\nabla^2 \vec{B} - \frac{1}{\delta_L^2} \vec{B} = 0, \tag{169}$$

with the *London penetration depth* given by,

$$\delta_L = \sqrt{\frac{m_e c^2}{4\pi e^2 n_s}} = \frac{c}{\omega_p}, \tag{170}$$

where ω_p is the plasma frequency in the material. Taken along one direction, say *z*, (170) becomes,

$$\frac{d^2 B_x}{dz^2} - \frac{1}{\delta_L^2} B_z = 0, \tag{171}$$

where $B_x(0)$ is the magnetic field at the surface of the superconductor. The solution stipulates that the magnetic field decays inside the superconductor with a characteristic length δ_L. Assuming a plasma frequency of 10^{15} /s, the approximate value of the London penetration depth is 300Å. This means that at distances greater than ~300Å from the surface, the magnetic field and, per (165), the current, vanish inside a superconductor, see Fig. 3-22.

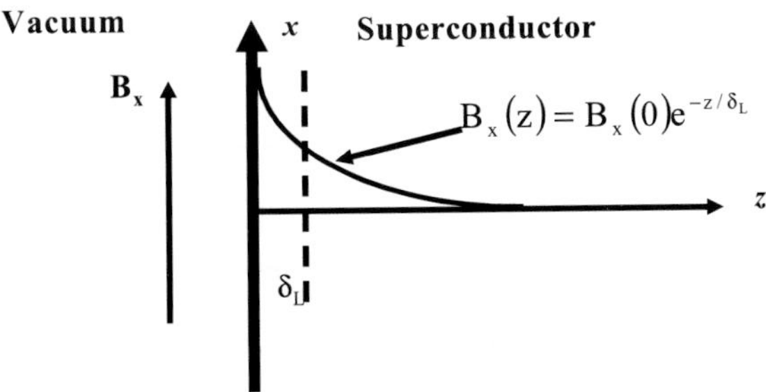

Figure 3-22. Decaying magnetic field in superconductor.

The vanishing of the magnetic field inside a superconductor is called the Meissner effect, and has certain practical consequences. For instance, if a superconducting wire is turned into a ring, then the fact that its bulk magnetic field and current are zero implies that,

$$\hbar\nabla\chi = \frac{2e}{c}\vec{A}' \Rightarrow \oint_C \nabla\chi \cdot \mathrm{dl} = \chi_2 - \chi_1 = \frac{2e}{\hbar c}\oint_C \vec{A}' \cdot dl \cdot \quad (172)$$

Therefore, the Cooper pair wave function may be written as,

$$\Psi_A = \Psi_0 e^{i\chi} = \Psi_0 e^{i\frac{2e}{\hbar c}\oint A\cdot dl} = \Psi_0 e^{i\frac{2e}{\hbar c}\phi}, \quad (173)$$

where ϕ is the magnetic flux inside the hollow part of the ring. Since the phase must equal an integer multiple of 2π, however, we have,

$$\frac{2e}{\hbar c}\phi = 2\pi n, \quad (174)$$

or,

$$\phi = \frac{\pi\hbar n}{ec} = \frac{1}{2}\left(\frac{2\pi hn}{e2\pi c}\right) = \frac{1}{2}\phi_0 n. \quad (175)$$

Thus, the magnetic flux confined by the superconducting ring is quantized in units of flux $\phi_0 = h/2e$, called a *fluxoid.*

The phase of the Cooper pair wave function and the fluxoid are at the heart of two effects of fundamental import for applications, namely, the Josephson effect and the nonlinear Josephson inductance.

The Josephson effect refers to the fact that, whenever two superconductors at the same temperature are brought in proximity to one another, separated by a thin insulating layer (so thin that tunneling of Cooper pairs may occur), Fig. 3-23, a supercurrent I_J flows, which depends on the

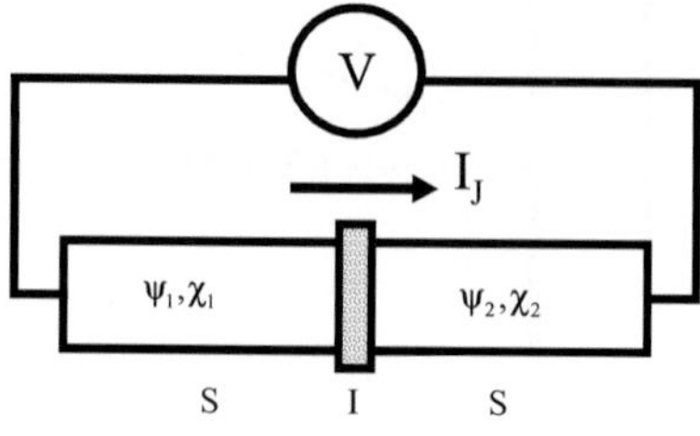

Figure 3-23. Schematic of Josephson junction.

phase difference $\delta = \chi_1 - \chi_2$ of the respective wave functions in the superconductors. Since the velocity of a Cooper pair is proportional to the phase gradient of its wave function, i.e., $v \sim \nabla\chi$, and since the phase has a period of 2π, it is not difficult to accept that the supercurrent be periodic. Indeed, it can be shown [28] that the Josephson junction current is given by,

$$I_J = I_0 \sin\delta, \tag{176}$$

where,

$$V = \frac{\phi_0}{2\pi}\frac{d\delta}{dt}, \tag{177}$$

is the voltage across the junction.

The Josephson inductance, in turn, derives from substituting (176) and (177) in the definition of inductance voltage, namely,

$$V = L_J \frac{dI_J}{dt}. \tag{178}$$

Thus,

$$\frac{dI_J}{dt} = I_0 \cos\delta \cdot \frac{d\delta}{dt} = I_0 \cos\delta \frac{2\pi}{\phi_0} V, \tag{179}$$

and, from (178) we obtain,

$$L_J = \frac{\phi_0}{2\pi I_0 \cos\delta}. \tag{179}$$

Clearly, the denominator, $\cos\delta$ makes the inductance nonlinear, becoming large as $\delta \to \pi/2$, and negative in the range $\pi/2 < \delta < 3\pi/2$. The nonlinearity of the Josephson inductance gives rise to the formation of the *Josephson qubit*, which is a nonlinear LC resonator consisting of the Josephson junction's inductance, L_J, and capacitance.

To conclude our exposition on superconductivity, we point out that there are two types of superconductors according to how the Meissner effect manifests in them [28]. In particular, type I superconductors are characterized by a magnetization versus applied magnetic field curve that increases up to a critical field, H_c, where it drops to zero and, concurrently, the superconducting state disappears (it becomes normal). Type II superconductors, on the other hand, are characterized by two critical fields, namely, a lower critical field H_{c1}, below which the superconducting state exists exclusively, and above which the superconductor is threaded by flux lines that give rise to a lattice of vortices, and an upper critical field H_{c2}, beyond which superconductivity disappears. The vortices are circulating superconducting currents around normal regions, and are such that the onset of a vortex occurs when the corresponding flux is that of a single fluxoid. Quantitatively,

$$H_{c1} \approx \frac{\phi_0}{\pi\delta_L^2}, \tag{180}$$

and

$$H_{c2} \approx \frac{\phi_0}{\pi\xi^2}, \tag{181}$$

where δ_L is the magnetic field penetration depth, and $\xi = \hbar v_F / 2\Delta$ [28] is the coherence length, which captures the lattice constant of vortex lattice.

3.2.2 Photonic Band-Gap Crystals

Continuing with the topic of wave phenomena in periodic structures, we now briefly take on the subject of electromagnetic wave propagation and manipulation in periodic dielectric structures or photonic band-gap crystals (PBCs) [51]. PBCs are 1-, 2-, or 3-dimensionally periodically patterned materials whose dispersion relation, i.e., propagation constant versus frequency response, exhibits ranges in which wave propagation is forbidden (band gaps) and ranges in which it is allowed.

3.2.2.1 One-dimensional PBC Physics

The fundamental physics of a PBC are easily grasped from considerations of a 1-D PBC, which is of finite extent and consists of alternating regions of dielectric constant, ε_1 and ε_2, respectively, see Fig. 3.24.

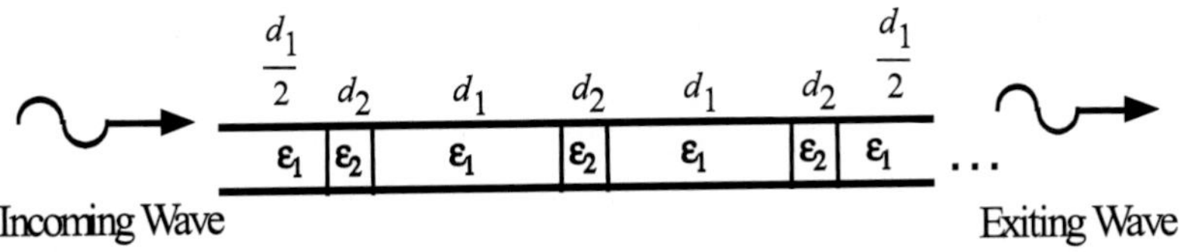

Figure 3-24 Sketch of one-dimensional PBC.

Focusing on a unit cell, see Fig. 3-25, we notice that if a wave impinges from the left on this unit cell, it will in general, undergo multiple reflections and trasmissions at two places, namely, *t, r* at the first $\varepsilon_1/\varepsilon_2$, discontinuity, and *r', t'* at the $\varepsilon_2/\varepsilon_1$ discontinuity.

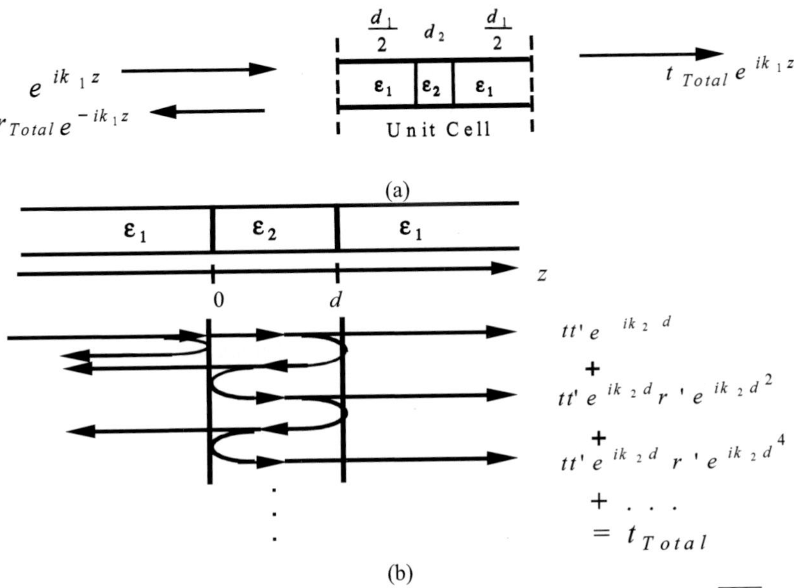

Figure 3-25 (a) PBC unit cell. (b) Transmission/reflection analysis. $k_i = \omega\sqrt{\mu\varepsilon_i}$ is the wave vector in region *i*.

Then, the amplitude of the transmitted wave will be given by the sum of the following terms [58]:

(1) The fraction that is transmitted through the $\varepsilon_1/\varepsilon_2$ interface, is phase-shifted while traversing (left-to-right) the region ε_2 of length *d*, and then is transmitted through the $\varepsilon_2/\varepsilon_1$ discontinuity, namely, $te^{ik2d}t'$. This is the amplitude for direct transmission through two discontinuities.

(2) The fraction that is reflected from $\varepsilon_2/\varepsilon_1$, is phase-shifted while traversing (right-to-left) ε_2 of length *d*, and then is reflected again at $\varepsilon_1/\varepsilon_2$, phase-shifted left-to-right the region ε_2 of length *d*, and so on. This is the amplitude for transmission after two reflections, and so on.

The frequency selectivity originates as follows [58]. At frequencies where *k2d* is an even multiple of $\pi/2$, we have,

$$t_{\text{Total}}\left(k_2 d = \text{even number} \cdot \frac{\pi}{2}\right) = tt'\left(1 + r'^2 + r'^4 + \ldots\right), \tag{182}$$

that is, every term inside the parenthesis is exactly in phase and there is constructive interference; this results in maximum transmission.

On the other hand, if *k2d* is an odd multiple of $\pi/2$, we have,

$$t_{\text{Total}}\left(k_2 d = \text{odd number} \cdot \frac{\pi}{2}\right) = tt'\left(1 - r'^2 + r'^4 + \ldots\right), \tag{183}$$

that is, every term inside the parenthesis alternates in sign and there is destructive interference, which results in minimum transmission. With $Z_i = \sqrt{\frac{\mu_i}{\varepsilon_i}}$ representing the characteristic impedance of region ε_i, we obtain the complex reflection and transmission coefficients as follows,

$$r' = \frac{Z_2 - Z_1}{Z_2 + Z_1} = \frac{\sqrt{\varepsilon_1} - \sqrt{\varepsilon_2}}{\sqrt{\varepsilon_1} + \sqrt{\varepsilon_2}}, \tag{184}$$

$$t' = \frac{2Z_2}{Z_2 + Z_1} = \frac{2\sqrt{\varepsilon_1}}{\sqrt{\varepsilon_1} + \sqrt{\varepsilon_2}}. \tag{185}$$

The real reflection and transmission coefficients are given by,

$$R = |r'|^2, \tag{186}$$

and

$$T = |t'|^2 \frac{Z_1}{Z_2}. \tag{187}$$

The overall transmission coefficient for the $\varepsilon_1/\varepsilon_2/\varepsilon_1$ of Fig. 3-25(b) is given by,

$$T_{Total} = |t_{Total}|^2 = \frac{T^2}{1 + R^2 - 2R\cos 2k_2 d}. \tag{188}$$

This expression can be used to compute the transmission coefficient of the unit cell, which includes *finite* ε_1 regions of length $d_1/2$, by replacing $k_2 d \rightarrow (k_1 d_1 + k_2 d_2)$. Figure 3-26 shows a plot of the transmission coefficient of such a unit cell, Eq.(188).

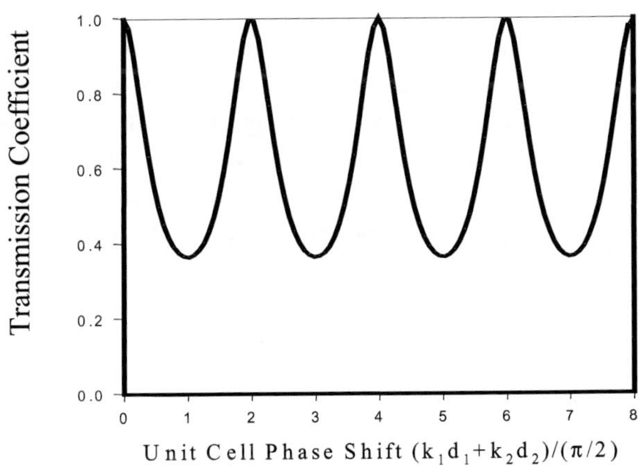

Figure 3-26. Transmission coefficient versus phase shift for unit cell for PBC in Fig. 3-25. Parameters: d_1=1.06in, d_2=0.42in, ε_1=1, ε_2=8.9. At odd multiples of $\pi/2$ one finds valleys, whereas at even multiples of $\pi/2$ one finds peaks of the transmission coefficient. The destructive interference, of a single unit cell in this example, is responsible for a valley transmission amplitude of only ~0.36. As the number of consecutive unit cells, N, making up the crystal increases, the cumulative effect of the unit cell's attenuation drives the overall crystal attenuation from ~0.36, for a single unit cell, to arbitrarily low values, depending on N. [161].

When multiple layers of unit cells are cascaded, the total transmission is drastically reduced and a photonic bandgap is formed at the frequency in question.

The 1-D PBC, being most often found in its embodiment as a multilayer film in dielectric mirrors and in optical filters, is already an extensively studied structure. From these applications it is known that PBCs can act as perfect mirrors for light whose frequency lies within a well-defined range, namely, when kd (where *d* is the lattice constant) is an odd multiple of $\pi/2$, and that they may localize modes when endowed with defects [162]. The application of PBCs in the context of routing and controlling the propagation of light waves, for example, requires their realization in, at least, 2-D. Next, we deal with multi-dimensional PBCs.

3.2.2.2 Multi-dimensional PBC Physics

The properties of 2- and 3-D PBCs may be formulated in terms of the coherent scattering properties of 2- and 3-D lattices [64]. Fig. 3-27(a) typifies a 2-D triangular-lattice PBC consisting of cylinders of dielectric constant ε_2 embedded in a host of dielectric constant ε_1.

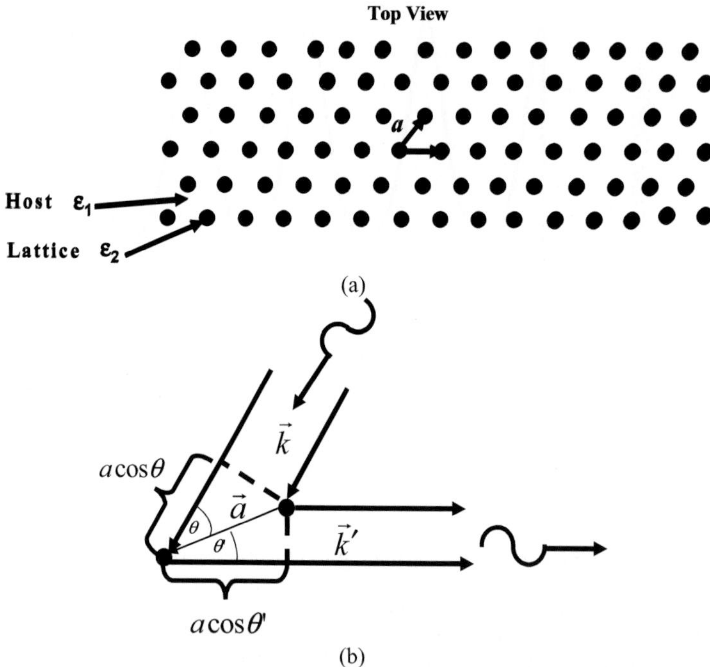

Figure 3-27. (a) Sketch of 2-D PBC with lattice constants $\vec{a}_1$ and $\vec{a}_2$ consisting of cylinders of dielectric constant ε_2 embedded in a host of dielectric constant ε_1. (b) Detail of an incoming wave with wave vector $\vec{k}$ impinging on two objects separated a distance $\vec{a}$, and scattered along wave vector $\vec{k}'$.

Specifically, the properties of these structures are given by arguments advanced by Bragg [64], whose essence (for a 2-D periodic lattice) is that the path difference (phase shift) between incoming and scattered rays, $\Delta = \vec{a} \cdot (\vec{k} - \vec{k}')$, see Fig. 3-27(b), determines whether the transmission of the structure exhibits a maximum or a minimum; a maximum when Δ is an integer multiple of 2π, and a minimum when it is an odd multiple of π. For a 3-D PBC, on the other hand, $\Delta = \vec{R} \cdot (\vec{k} - \vec{k}')$ must be valid simultaneously for all vectors $\vec{R}$ that are Bravais lattice vectors [64].

A large number of computational techniques to obtain the properties of general PBCs have been developed, most of which derive from the solid state physics literature on computing band structures [162]-[166]. Obviously, it would be impossible to engage in detailing these techniques here, thus we instead provide a number of analytical results derived by Joannopoulos *et al.* [162] that capture some general properties of PBCs and facilitate one's intuition when thinking about them.

3.2.2.2.1 General Properties of PBCs

Initially, techniques for computing the properties of dielectric PBCs exploited previously introduced methods for computing the band structures of semiconductors. Indeed, a comparison between the equations of quantum mechanics (QM), used to describe semiconductors, and electromagnetics (EM), used to describe dielectric PBCs, shows many similarities, Table 3-1.

Table 3-1. Comparison between quantum mechanics and electromagnetics formulations. [159].

Field	$\Psi(\vec{r},t) = \Psi(\vec{r})e^{i\omega t}$	$\vec{H}(\vec{r},t) = \vec{H}(\vec{r})e^{i\omega t}$
Eigenvalue problem	$H\Psi = E\Psi$	$\Xi\vec{H} = (\omega/c)^2\vec{H}$
Hermitian operator	$H = \dfrac{-(\hbar^2\nabla^2)}{2m} + V(\vec{r})$	$\Xi = \nabla\times\left(\dfrac{1}{\varepsilon(\vec{r})}\nabla\times\right)$

A key difference, however, which restricts the general applicability of the QM formulation to solve PBC problems is the scalar nature of the QM problem compared to the vector nature of the EM problem. Fortunately, however, unlike the QM semiconductor band structure problem, in which the Bohr radius introduces a fundamental length scale and, as a result, similar lattices with differing dimensions give rise to different behaviors, the EM problem possesses no fundamental length scale constant. This means that the properties of PBCs which differ only via a length expansion or contraction of all distances, are related by simple expressions. In particular, given an EM eigenmode obeying the equation,

$$\nabla\times\left(\frac{1}{\varepsilon(\vec{r})}\nabla\times\right)\vec{H}(\vec{r})=\left(\frac{\omega}{c}\right)^2\vec{H}(\vec{r}), \tag{189}$$

if the dielectric profile defining a PBC is scaled as follows, $\varepsilon(\vec{r})\rightarrow\varepsilon'(\vec{r})=\varepsilon(\vec{r})/s$, where *s* is the scaling factor, then it can be shown that the scaled PBC will obey the equation,

$$\nabla\times\left(\frac{1}{\varepsilon'(\vec{r}')}\nabla'\times\right)\vec{H}(\vec{r}'/s)=\left(\frac{\omega}{cs}\right)^2\vec{H}(\vec{r}'/s), \tag{190}$$

from where one derives that the properties corresponding to the scaled PBC are derived from those of the unscaled one as follows: $\vec{H}'(\vec{r}')=\vec{H}(\vec{r}'/s)$ and $\omega'=\omega/s$. Thus, once the PBC solutions are known at one length scale, they are automatically known at all others. As a practical application, microwave-length-scale PBCs may be exploited as vehicles to study to optical-scale PBC concepts.

Similarly, there is no fundamental value of dielectric constant, therefore, it may be shown that whenever the dielectric constant is uniformly scaled throughout a PBC as follows: $\varepsilon(\vec{r})\rightarrow\varepsilon'(\vec{r})=\varepsilon(\vec{r})/s^2$, where *s* is the scaling factor, then the scaled PBC will obey the equation,

$$\nabla\times\left(\frac{1}{\varepsilon'(\vec{r})}\nabla\times\right)\vec{H}(\vec{r})=\left(\frac{s\omega}{c}\right)^2\vec{H}(\vec{r})\cdot \tag{191}$$

This means that, upon scaling the dielectric constant, the mode geometry remains unchanged, but the frequency scales as: $\omega\rightarrow\omega'=s\omega$. Thus, multiplying the dielectric constant by a factor of 1/9 will result in multiplying the frequency of their modes by three.

Lastly, the properties of PBCs depend on parameters such as filling fraction, the contrast between host and lattice dielectric constants, and the number of layers employed. Fig. 3-28 shows the computed transmission coefficient for an eleven-layer PBC as the index of refraction $n=\sqrt{\varepsilon_2}$ is increased from 1.2 to 2.98.

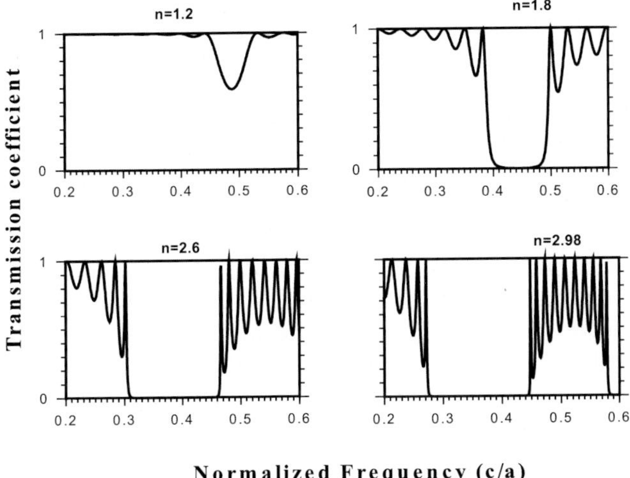

Figure 3-28. Eleven-layer 2-D PBC transmission coefficient with index of refraction as a parameter. The band gap attenuation increases from a few dB for n=1.2 to close to 80dB at n=2.8.

3.2.2.3 Advanced PBC Structures

The initial investigations in the field of PBCs focused on dielectric materials-based PBCs, whose structure consisted of either periodic arrays of suitably shaped holes in a dielectric slab, thus forming a continuous dielectric host matrix, or a periodic array of suitably shaped and isolated dielectric objects. The former PBC is exemplified by a slab patterned with an array of cylindrical air holes, whereas the latter PBC is exemplified by an array of isolated cylinders embedded in air. These PBCs permitted the creation of band gaps at finite frequencies, but did not produce them at DC. Further investigations on metal-based PBCs, such as a wire mesh, soon followed, which demonstrated the existence of band gaps down to DC [167], [168].

While enabling the manipulation of electromagnetic waves, in particular, achieving diffractionless guidance of light around sharp bends [162], the overall propagation behavior in dielectrics and metallic meshes followed the usual "right-hand" (RH) rule, in which the directions of the electric and magnetic fields, $\vec{E}$ and $\vec{H}$, and the propagation vector $\vec{k}$ form a right-handed system with coincidence of the direction of energy flow and $\vec{k}$. Further work, aimed at manipulating the properties of the PBC medium, led Pendry to propose two schemes, namely, a composite medium made up of an array of metal posts which created a frequency region with negative permittivity, $\varepsilon_{\text{eff}} < 0$, and an array of interspersed split-ring resonators which created a frequency region with negative permeability, $\mu_{\text{eff}} < 0$. These materials have become known as *metamaterials* and, when implemented so that both the permittivity and the permeability are simultaneously negative, they exhibit a negative refractive index $n(\omega) = \sqrt{\varepsilon_{\text{eff}}(\omega)\mu_{\text{eff}}(\omega)}$, which is real and gives rise to the existence of propagating modes with the remarkable property that they follow a "left-hand" (LH) rule. In this case the vectors $\vec{E}$, $\vec{H}$, $\vec{k}$ form a left-handed system, i.e., the direction of propagation is reversed with respect to the direction of energy flow [169]. Left-handed materials have been the subject of much attention because they exhibit unusual propagation properties. For instance, they exactly reverse the propagation paths of rays within them, which may be exploited to implement low reflectance surfaces by exactly canceling the scattering properties of other materials. Another application, exploits their potential to produce perfect lenses.

3.2.2.3.1 Negative Refraction and Perfect Lenses

The concept of a perfect lens was introduced by Pendry [170], upon further examining the earlier analysis of Veselago [169] on the consequences of negative refractive index materials. Veselago [169], in particular, had indicated that reflection and refraction between vacuum and a negative refraction material, would follow the situation depicted in Fig. 3-29.

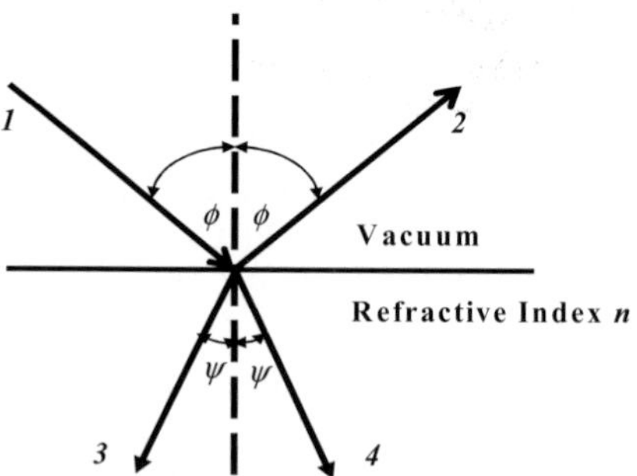

Figure 3-29. Consequences of negative refractive index on refraction properties. 1—Incident beam. 2—Reflected beam. 3—Refracted beam for $n<0$. 4—Refracted beam for $n>0$. (*After* [169].)

Fig. 3-29 shows, that contrary to the usual case of a positive index, when the refraction index is negative the angle of refraction is also negative with respect to the surface normal. As a result, when such a medium is used as a lens, Fig. 3.30, it causes light originally diverging from a point source S to be reversed and to converge back to a point S_1 in the medium.

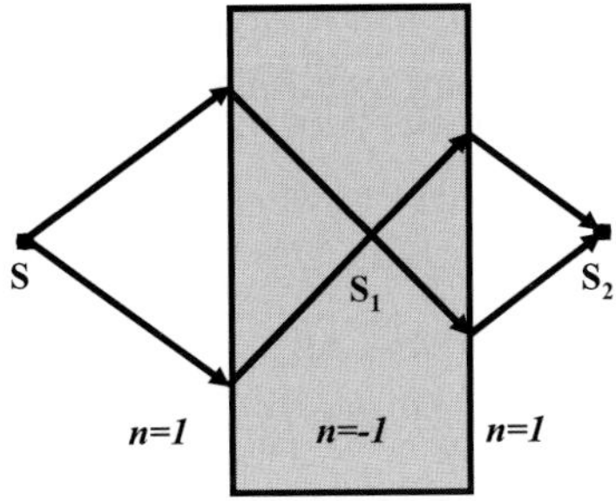

Figure 3-30. Parallel-sided medium with negative refractive index refocuses light. (*After* [170].)

The special feature contributed by a negative refraction lens was elucidated by Pendry [170]. It consists in that, by being capable of amplifying evanescent waves, all wave components emanating from the source are present at the converging focus; this enables the perfect reconstruction of the source image. This property was proven by Pendry [170] by assuming an incident wave with electric field given by,

$$E_{0S+} = \hat{y}\exp(ik_z z + ik_x x - i\omega t), \tag{192}$$

where, since $k_x^2 + k_y^2 > \omega^2/c^2$, the wave vector,

$$k_z = +i\sqrt{k_x^2 + k_y^2 - \frac{\omega^2}{c^2}}, \tag{193}$$

implies an exponentially decaying (evanescent) wave, a reflected wave given by,

$$E_{0S-} = r\hat{y}\exp(-ik_z z + ik_x x - i\omega t), \tag{194}$$

and transmitted wave given by,

$$E_{1S+} = t\hat{y}\exp(ik'_z z + ik_x x - i\omega t), \tag{195}$$

where,

$$k'_z = +i\sqrt{k_x^2 + k_y^2 - \frac{\varepsilon\mu\omega^2}{c^2}} \quad ,k_x^2 + k_y^2 > \frac{\varepsilon\mu\omega^2}{c^2}. \tag{196}$$

Then, using the formula for the transmission coefficient of a slab of width d, i.e.,

$$T_S = \frac{tt'\exp(ik'_z d)}{1 - r'^2 \exp(2ik'_z d)}, \tag{197}$$

where,

t and t' are the vacuum/medium and medium/vacuum transmission coefficients and r and r' the corresponding reflection coefficients, given by,

$$t = \frac{2\mu k_z}{\mu k_z + k_z'} \quad ,r = \frac{\mu k_z - k_z'}{\mu k_z + k_z'}, \tag{198}$$

and

$$t' = \frac{2k'_z}{k'_z + \mu k_z} \quad ,r' = \frac{k'_z - \mu k_z'}{k'_z + \mu k_z}. \tag{199}$$

If both the permeability and permittivity approach negative unity, then the transmission coefficient becomes,

$$\lim_{\substack{\mu \to -1 \\ \varepsilon \to -1}} T_S = \lim_{\substack{\mu \to -1 \\ \varepsilon \to -1}} \frac{tt'\exp(ik'_z d)}{1 - r'^2 \exp(2ik'_z d)} = \exp(-ik'_z d) = \exp(-ik_z d)\cdot \tag{200}$$

Since k_z is imaginary, see (196), (200) is a growing exponential and the wave is amplified.

By contrast, in a normal lens the large transverse wave vector of propagating waves are evanescent and decay prior to reaching the focus, thus the incomplete spectral contents makes it impossible to identically reconstruct the image.

3.2.3 Cavity Quantum Electrodynamics

The field of Cavity Quantum Electrodynamics or, cavity QED, deals with the effect of the surrounding environment on the spontaneous emission rate of atoms [171]. The concept was introduced by Purcell in 1946 [171] in the context of nuclear magnetic moment transitions. He observed that at conditions of temperature, radio frequency, and nuclear magneton given by 300°K, $\nu = 10^7\ \mathrm{sec}^{-1}$, and $\mu = 1$, respectively, the corresponding rate of spontaneous emission, given by,

$$A_\nu = \left(\frac{8\pi\nu^2}{c^2}\right) h\nu \left(\frac{8\pi^3\mu^2}{3h^2}\right) \ \mathrm{sec}^{-1}, \tag{201}$$

adopts a value of $2\times10^{-22}\ \mathrm{sec}^{-1}$. So small is, indeed, this value, that it implies the virtual impossibility of the spin system being able to achieve thermal equilibrium with its surroundings. This expression, Eq. (201), for the spontaneous emission rate A between initial and final states $|i\rangle$ and $|f\rangle$, assumes the atom is in free space and derives from Fermi's golden rule [172], namely,

$$A = \frac{|\langle f|H|i\rangle|^2}{\hbar^2}\rho(\nu), \tag{202}$$

where the initial state $|i\rangle$, represents an atom in the absence of any photons, and the final state $|f\rangle$, represents the atom with a single photon. The Hamiltonian H represents the atom-field interaction, and $\rho(\nu)$ represents the density of photon states or number of radiation oscillators per unit volume, in a unit frequency range which, for free space, adopts the value of ,

$$\rho_S = \left(8\pi\nu^2/c^3\right). \tag{203}$$

In other words, ρ_S embodies the number of electromagnetic modes into which photons may be emittted at the location of the emitter [173].

When the atom is enclosed by a microwave cavity of quality factor Q, however, the number of radiation oscillators per unit volume is limited to those occupying the frequency range ν/Q, which is, in fact, exactly one. If one assumes the cavity volume and the wavelength to be related by

$V \cong (\lambda/2)^3 = (c/2\nu)^3$, then the density of photon states per unit frequency, per unit volume, $(1\ \ mode)/(\Delta\nu V)$, may be expressed in terms of the cavity Q as follows,

$$\frac{1 \text{ mod e}}{(\Delta\nu)(V)} = \frac{1 \text{ mod e}}{\left(\frac{\nu}{Q}\right)\left(\frac{c}{2\nu}\right)^3} = \frac{8\nu^2 Q}{c^3} = \rho_c \,. \tag{204}$$

Comparing (203) and (204) it is seen that they are related by,

$$\rho_c \cong \left(\frac{2}{\pi}\right) Q \cdot \rho_s \,. \tag{205}$$

Thus, a cavity enclosure of quality factor Q increases the effective density of photon states in free space by the factor of $(2Q/\pi)$. In turn, since the spontaneous emission rate is proportional to this density of photon states, this rate is increased, in particular, to [172],

$$A_c \cong QA \,. \tag{206}$$

The larger issue elicited by Purcell's observation was that the spontaneous emission rate of an atom may be modified according to the properties of the surroundings. In particular, as Kleppner [172] pointed out, the spontaneous emission of an atom in a cavity may be inhibited if the cavity has dimensions smaller than the radiaton wavelength, but it may be enhanced (increased), as in (206), if the cavity resonates at this wavelength.

This realization that the spontaneous emission rate of an atom may be suppressed or enhanced *by modifying the properties of the radiation field in the surroundings*, has many practical applications. For instance, in solid-state electronics it is well known that spontaneous emission is fundamentally responsible for non-radiative recombination processes, which limit the performance of semiconductor lasers, heterojunction bipolar transistors, and solar cells [51]. How would one apply the cavity QED concept to inhibit the spontaneous emission in these situations, where one is dealing not with single atoms, but with entire devices, is not at all obvious. The answer to this question was advanced by Yablonovitch in 1987 [51] with his photonic band-gap crystal (PBC) idea. Indeed, by surrounding the devices in question with a PBC exhibiting a band gap which overlaps the electronic band edge (across which the non-radiative transitions would occur) the spontaneous emission can be forbidden, thus potentially eliminating non-radiative transitions. This is so because, in the band gap of a PBC, the density of photons states, $\rho_{PBC} = 0$. The first experimental demonstration of the use of three-dimensional PBCs to control the dynamics of spontaneous emission from quantum dots has been recently published [173]. In this case, Fig. 3-31,

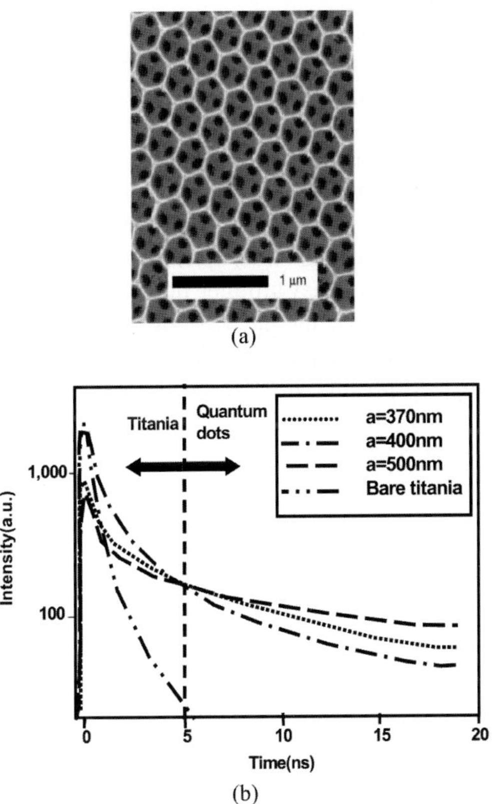

Figure 3-31. (a) Scanning electron microscope image of the (111) face of a titania inverse opal with lattice parameter a=460 nm. Reprinted with permission from[170] Copyright 2004 *Nature*. (b) Luminescence decay curves of quantum dots inside three different photonic crystals. The data are recorded at frequencies 15,670 cm^{-1} (a= 370 nm) and 15,100 cm^{-1} (a =420 nm, and a=500 nm). The curves have been overlapped after 5 ns. The first part of the decay curve is influenced by emission of titania (recorded at 15,400 cm^{-1}). After 5 ns this contribution is negligible. [173].

the spectral distribution and time-dependent decay of light emitted from excitons confined in the CdSe quantum dots are shown to be controlled by the host PBC. In particular, the fact that lifetimes of 9.6 ± 0.1ns and 19.3 ± 0.2ns for quantum dots embedded in PBCs of lattice constants a=420 nm and a=500 nm, respectively, are obtained, demonstrate a factor of

2 variation produced by the PBCs. This corroborates the strong role the PBC plays in controlling the radiative lifetime of the emitters.

3.3 Summary

This chapter has dealt with the physics of waves that is of relevance to quantum phenomena occurring in NanoMEMS. It began with typical phenomena that manifest and exploit the wave nature of electrons, in particular, the quantization of electrical conductance, its calculation with Landauer's formula, and its manifestation in quantum wires, quantum point contacts, resonant tunneling and quantum interference (Aharonov-Bohm effect). Then, the topic of quantum transport theory was taken up, with particular emphasis on dealing with phenomena dominating in one-dimensional transport, such as the Lüttinger liquid. Finally, wave behavior in both periodic and non-periodic media was addressed, in particular, carbon nanotubes, superconductors, photonic bandgap crystals, and cavity quantum electrodynamics. In next chapter focuses on the application of the material presented thus far to engineer a variety of circuits and systems that typify elements to be found in NanoMEMS.

Chapter 4

NANOMEMS APPLICATIONS: CIRCUITS AND SYSTEMS

4.1 Introduction

The new "electronics," enabled by NanoMEMS, will exploit the coexistence of mesoscopic and mechanical devices operating in the *quantum mechanical* regime. Thus, a plethora of phenomena, such as tunneling, charge quantization, the Casimir effect, *motion* quantization, entanglement, etc., are at our disposal to be exploited in creating powerful computing and communications hardware. This chapter exposes a variety of emerging devices that embody potential nanoelectromechanical quantum circuits and systems (NEMX) device-circuit paradigms [22].

4.2 NanoMEMS Systems on Chip

NanoMEMS Systems-on-Chip (SoC) may be predicated upon a multitude of physical phenomena, e.g., electrical, optical, mechanical, magnetic, fluidic, quantum effects and mixed domain. Therefore, their universe of possible implementations and applications is vast and only limited by our imagination. Possible areas of endeavor, already under research, include: Nanoelectronics, Nanocomputation, Nanomechanics, Nanoengineering, Nanobiotechnology, Nanomedicine, Nanochemistry, and RF MEMS. In principle, then, there is the potential for conceiving new devices that might spark a revolution as important and wide-ranging as that engendered by the invention of the transistor and ICs. Ultimately, however, the success of the technology may well lie on its ability to deliver improved performance at low cost on *technology-blind* applications, Figure 4-1, as well as in enabling new applications (some of which are right now only limited by our imagination). For the purposes of this book, we focus on NanoMEMS SoCs in terms of implementation and applications.

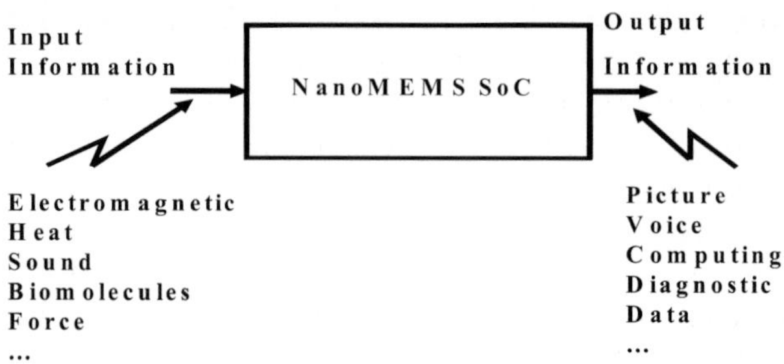

Figure 4-1. Conceptual rendition of a NanoMEMS System-on-Chip.

4.2.1 NanoMEMS SoC Architectures

Regardless of the technology of implementation utilized, a system must perform a definite function and is characterized by how close it comes to meeting certain *technology-blind* specifications (specs). Typically, the design process begins with a block diagram of the *system* in question, which displays an *architecture* or high-level topological diagram showing how the constituent *building blocks* are interconnected to transform or *process* one or more input signals into one or more output signals, see Figure 3-1. Following this, overall systems analysis assigns or "flows down" the overall system specs to the individual building blocks, which are then designed. In the case of NanoMEMS SoCs this is difficult to do because the field is so premature that, using a circuit analogy, the equivalents of passive components (resistors, inductors, capacitors, diodes) and active components (transistors) is not yet available to the degree of completeness that would allow a complete consistent system implementation. Our course of action, therefore, is to expose a variety of potential NanoMEMS SoC building blocks.

4.2.2 NanoMEMS SoC Building Blocks

4.2.2.1 Interfaces

The idea behind NanoMEMS is that of creating a system that, in order to accomplish a given function, avails itself of devices and techniques spanning the range from the *micro*- down to the *nano*-scale and beyond. In the most general case, the input signal to a NanoMEMS SoC will be *analog*, i.e., will exhibit continuous amplitude and will exist at all times, see Figure 3-1. Processing this signal, therefore, will entail deciding whether it is feasible to act on it as received/detected, or to transform it to a more convenient state. The nature of the interface *sensor*, in particular, its sensitivity, bandwidth, and dynamic range, will come into play here and will dictate the need for transduction, amplification, digitization, filtering, etc., thus determining the rest of the architecture. In this context, the doubly-anchored Si beam has been considered as a potential mechanical sensing element in future NanoMEMS SoCs, and impressive estimates for its *intrinsic* force sensitivity (S_F), dynamic range (DR), mass sensitivity (M), and bandwidth (BW) have been obtained by Roukes [174]. For instance, a beam of length, width, and thickness 0.1 x 0.01 x 0.01 microns and active mass 10ag would exhibit $S_F^{1/2}(\omega_0) = 3\times10^{-17} N/\sqrt{Hz}$, $DR = 35dB$, $M = 1.7\times10^{-21} g$, and $BW = 7.7GHz$, assuming a temperature of 300K and a Q of 10,000. Unfortunately, it is unclear whether the full extent of these parameters will be accessible due to various practical difficulties such as mass variation due to unpredictable adsorbates, and the impossibility of realizing a noiseless read-out. This latter theme is also common to electrostatic- and optically-based sensing interfaces as well. In the former case, which according to Roukes [174] may attain a minimum capacitance of $10^{-18} F$, the parasitic capacitance would preclude resolving it. In the latter case, the fact that the spot size of the light delivered by the optical fiber used in AFM displacement-sensing is much greater than nanoscale dimensions, precludes its resolution and, hence, proper detection.

In systems with an electronic input signal sensing scheme, however, the sensor may take the form of a quantum superlattice-based analog-to-digital converter, Fig. 4-2 [175]. Here, the pulsating nature of the superlattice's current-voltage characteristic directly samples/quantizes the voltage axis. The resulting current is used to generate pulses that drive a counter whose output is a digital representation of the input voltage. For highest resolution, the superlattice may be realized with molecular devices.

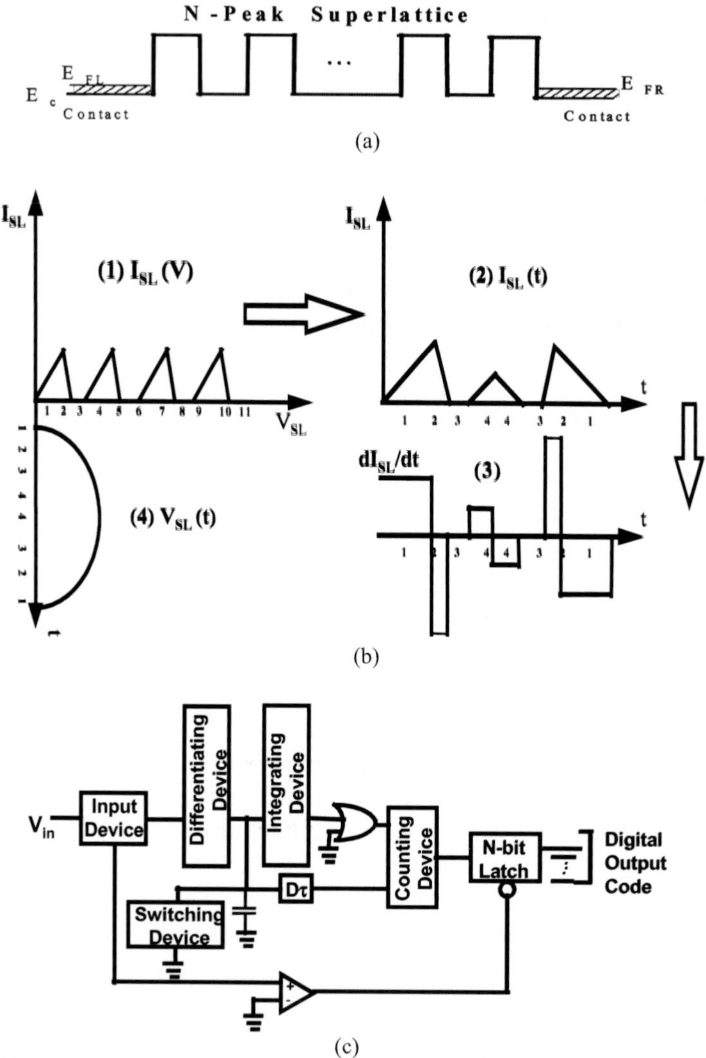

Figure 4-2. Superlattice-based analog-to-digital converter architecture. (a) Superlattice band diagram. (b) SL A/D conversion principle. (c) ADC architecture [175].

4.2.2.2 Emerging Signal Processing Building Blocks

While the specific structure of a NanoMEMS SoC is still the subject of much research, a number of potential building blocks for NanoMEMS-based signal processing have been proposed. In what follows, we present a number of these [22], namely, a charge detector, a which-path electron interferometer, a parametric amplifier using a torsional MEM resonator, a

Casimir effect-based oscillator, a magnetomechanically actuated beam, and array-based functions. We conclude with an example of exploiting quantum squeezing to reduce noise in mechanical structures.

4.2.2.2.1 Charge Detector

This device was experimentally demonstrated by Krömmer *et al.* 176]. In this device a low-power RF signal propagates through a suspended resonator, Figure 4-3, and sets it into vibration.

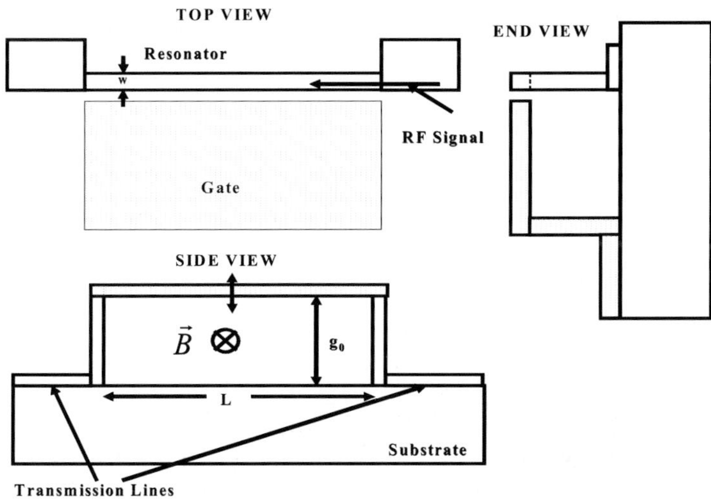

Figure 4-3. Schematic of charge detection resonator system [22].

With an in-plane magnetic field applied perpendicular to beam, a Lorentz force perpendicular to the substrate surface is developed. Application of a voltage, V, between the gate and the beam, induces a charge, Q, on the beam via the relation, $Q = CV$, and essentially, modifies its stiffness (spring constant). This results in a mechanical resonance frequency shift of $\delta f = \frac{Q^2}{2C}\left(1 - \frac{C'' z^2}{2C}\right)$, where C is the gate-beam coupling capacitance, and C'' represents the second derivative of the capacitance with respect to beam elongation amplitude, *z(t)*, evaluated at *z=0*. Optimum charge detection (maximum frequency shift) is obtained when RF power drives the beam to the verge of nonlinear amplitude vibration. For a gate bias of $V = \pm 4V$, a magnetic field of 12T, and an RF power of -52.8dBm at 37.29MHz, a charge detection resolution of about $70q/\sqrt{Hz}$. This device has the potential to exploit charge discreteness effect.

4.2.2.2.2 Which-Path Electron Interferometer

Armour and Blencowe [177], [178] presented a theoretical analysis for this concept. A cantilever resonator operating at radio frequencies is disposed over one of the arms of an Aharonov-Bohm (AB) [125] ring containing a quantum dot (QD), Figure 3-4. Electrostatic coupling of the vibrating beam with

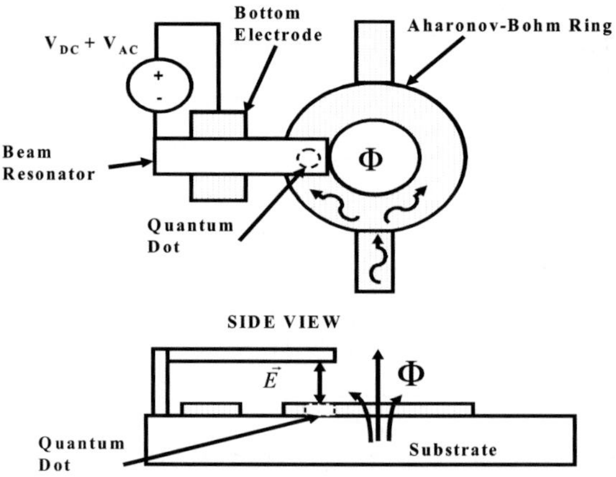

Figure 4-4. Schematic of mechanical which-path electron interferometer [22].

electrons hopping in/out of the QD modulates the interference fringes, according to vibration frequency (ω_0)-electron dwell time, $\tau_d = \hbar/\Delta E_{inc}$, product, where ΔE_{inc} is the electron energy spread. For $\omega_0 \tau_d << 1$, short dwell time, interference fringes are destroyed if $qE\Delta x_{th} > \Delta E_{inc}$., where x_{th} is the thermal position uncertainty of the cantilever and E the electric field. This signals electron dephasing and detection in QD arm. For $\omega_0 \tau_d \sim 1$, the beam-QD behaves as a coherent quantum system, beam vibration and QD exchange virtual energy quanta in resonance, and interference fringes are modulated at beam vibrating frequency. For the largest dwell times, the environment induces lost of coherence. This device has the potential to exploit charge discreteness effect.

4.2.2.2.3 Parametric Amplification in Torsional MEM Resonator

This device was experimentally demonstrated by Carr, Evoy, Sekaric, Craighead and Parpia [179]. A torsional resonator of quality factor Q, Figure 3-5, is excited at a fundamental driving frequency, ω, which applies

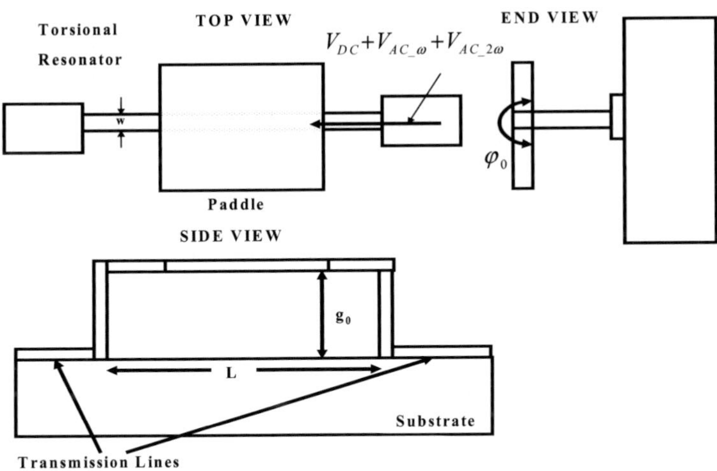

Figure 4-5. Schematic of torsional parametric amplifier [22].

a torque $\tau(\omega t)$. If the device is driven at resonance, with an applied torque given by $\tau(t) = \tau_0 \sin(\omega t + \theta)$, where θ is the phase angle between excitations at ω and 2ω, then the torsional spring constant exhibits a modulation, $\kappa'(t) = \kappa'_0 \cos(2\omega_0 t)$. Under these circumstances, the angular amplitude response, φ_0, adopts the form

$$\varphi_0 = \frac{\tau_0 Q}{\kappa}\left[\frac{\cos^2\theta}{(1+Q\kappa'_0/2\kappa)^2} + \frac{\sin^2\theta}{(1-Q\kappa'_0/2\kappa)^2}\right]^{1/2}. \qquad (1)$$

Accordingly, with zero signal amplitude at 2ω, $\kappa'(t) = 0$, and the angular response is $\tau_0 Q/\kappa$. Otherwise, the square-root factor acts as a phase-dependent gain and, becoming infinity when $\theta = \pi/2$, and $\kappa'_0 = 2\kappa/Q$. For $0 < \theta < \pi/2$, the angular response may be approximated by,

$$\varphi_0 = \frac{\tau_0 Q}{\kappa}\left[\frac{\cos^2\theta}{\left(1+V_{AC\_2\omega}/V'\right)^2}+\frac{\sin^2\theta}{\left(1-V_{AC\_2\omega}/V'\right)^2}\right]^{1/2}, \tag{2}$$

where V' is a structure-dependent parameter, showing that the gain increases with the pump signal amplitude. The device has the potential to exploit the Casimir effect.

4.2.2.2.4 Casimir Effect Oscillator

This device, which was proposed and analyzed by Serry, Walliser, and Maclay [180] in 1995, Figure 4-6, and experimentally realized by Chan, Aksyuk, Kleiman, Bishop and Capasso [181] in 2001, represents the first clear demonstration of the impact of the Casimir force in the performance of NEMX.

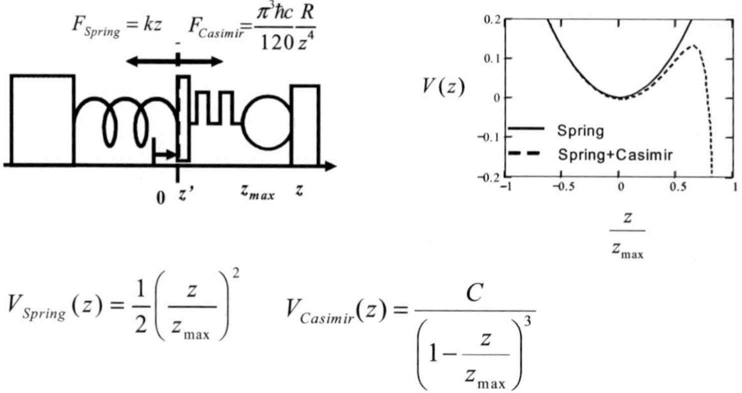

Figure 4-6. Summary of nonlinear Casimir effect MEM resonator physics [22].

The experiment entailed changing the proximity of a vibrating rotational resonator to a metallic sphere, Figure 4-7(a), to measure its behavior in the absence/presence of the Casimir force. After determining the drive for linear response, the proximity of the oscillator to a metallic sphere was varied and the resonance frequency measured exhibited a behavior as depicted in Figure 4-7(b). For sphere-oscillator distances greater than $3.3\mu m$, the oscillator resonance frequency was equal to the drive frequency, 2748Hz, and the angular amplitude frequency response was symmetric and centered around the drive frequency, $\omega_0 = \sqrt{k/I}$, where *k* is the spring constant and *I* the moment of inertia, consistent with mass-spring force oscillator behavior.

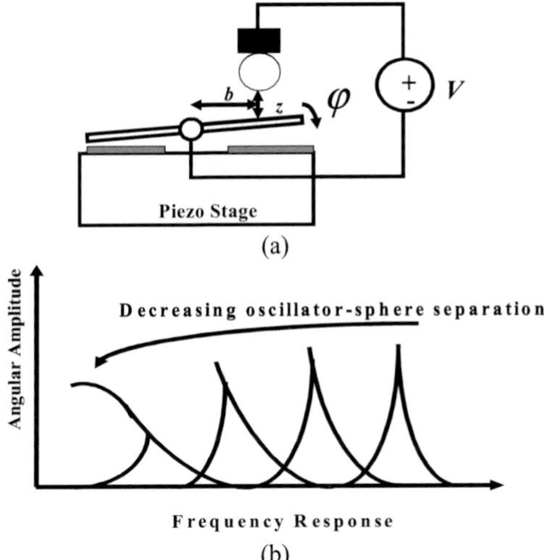

Figure 4-7. Schematic of torsional MEMS oscillator and sketch of Casimir effect on resonance response [22].

However, as the sphere-oscillator distance was decreased, in particular, at 141nm, 116.5nm, and 98nm, the resonance frequency shifted, according to, $\omega_1 = \omega_0 \left[1 - b^2 F'(z)/2I\omega_0^2\right]$, where $F'(z)$ is the first derivative of the external force evaluated at z, and the angular amplitude frequency response asymmetric and hysteretic. This behavior was shown to be consistent with the dynamics of a mass-spring-Casimir force system. The ramifications of this beautiful experiment are enormous, in particular, it may be concluded that the Casimir force will be one of the factors limiting the integration level or *density* of NEMX.

4.2.2.2.5 Magnetomechanically Actuated Beams

This idea was proposed and theoretically analyzed by Blom [182]. In addition to their function as mechanical elements (actuators), narrow *metal-coated* nanoscale beams also embody mesoscopic wires. If such a beam is elongated due to, say, electrostatic actuation, this results in a *reduction* in its cross-sectional area, and in particular, that of the current-carrying metallization/wire, and as a consequence, the conductance of the latter changes as transverse quantized modes are pushed above the Fermi level. The change in thermodynamic potential as the wire elongates, in turn produces a force along the length of the wire, which is given by,

$$F = \sum_n \sqrt{\frac{2m}{\pi^2 \hbar^2}} \left\{ \frac{4}{3} \left(E_F - E_n^c \right)^{3/2} - 2\left(E_F - E_n^c \right)^{1/2} E_n^c \right\}, \qquad (3)$$

where E_F is the Fermi energy, m the electron mass, and E_n^c is the energy of the transverse modes. This force manifests as force and beam deflection fluctuations. On the other hand, if the beam is not electrostatically actuated, but a magnetic field is applied along its length, it will also cause conductance changes as the Landau levels [60] push the energy above the Fermi level. Thus, the beam is magnetomechanically actuated. This devices has the potential to exploit charge discreteness effect.

4.2.2.2.6 Systems—Functional Arrays

The dynamic properties of the collective modes in a MEMS resonator array were studied experimentally by Buks and Roukes [183], and theoretically by Lifshitz and Cross [184]. In this concept, the lateral electrostatic coupling of an array of doubly-anchored beams leads to collective modes that resemble phonons. Adjustment of the coupling serves to *tune* the diffraction properties of the mechanical lattice the array embodies. In a related concept, De Los Santos [185] unveiled the idea of populating a rigid photonic band-gap crystal lattice with a sub-array of MEMS switches. Then, by exploiting the noninvasive properties of these, i.e., their ideal ON/OFF states, localized states modes could be formed that enabled the ON/OFF switching of pass bands within the photonic band-gap, thus making the system programmable.

4.2.2.2.7 Noise—Quantum Squeezing

Ultimately, the purity of resonator vibration is determined by its zero-point fluctuations. In this context, quantum squeezing techniques [186] may be applied to reduce the fluctuations in flexural motion. Application of quantum squeezing to *mechanical* resonators has been studied theoretically by Blencowe and Wybourne [187]. Accordingly, by exciting the resonator with a pumping voltage of the form, $V_p(t) = V_0 \cos(\omega_p t + \phi)$, its spring constant becomes, $k_0 = m\omega_0^2 + \Delta k$, where $\Delta k = C_0 V_0^2 / 2g_0^2$, and $k_p(t) = \Delta k \cos(2\omega_p t + 2\phi)$. When the effective resonator spring constant, $k = k_0 + k_p(t)$, increases, the curvature of the effective potential narrows [187] and this squeezes the wavefunction. In particular, for a phase $\phi = -\pi/4$, the quantum uncertainty in the flexural displacement becomes,

$$\Delta Z_1^2 \approx \frac{\hbar(2n_T + 1)}{2m\omega_0}\left(1 + \frac{Q\Delta k}{2m\omega_0^2}\right)^{-1}, \tag{4}$$

where $n_T = \left(e^{\hbar\omega/k_B T} - 1\right)^{-1}$. Then, with $\sqrt{\hbar/2m\omega_0}$ defining the zero-point uncertainty, the *squeezing factor* $R = \Delta Z_1 / \sqrt{\hbar/2m\omega_0}$ becomes,

$$R = \sqrt{\frac{2n_T + 1}{1 + Q\Delta k / 2m\omega_0^2}} < 1, \tag{5}$$

which, for $R < 1$, denotes the occurrence of *quantum* squeezing. Blencowe and Wybourne [174] found that using typical resonator values, e.g., density, $\rho = 3.99 \times 10^3\, kg/m^3$, Young's modulus, $E = 3.7 \times 10^{11}\, N/m^2$, beam to substrate distance, $g_0 = 50nm$, beam thickness, $t = 100nm$, and length, $L = 2700nm$, the squeezing factor is $R \approx 0.25$, which signals the realization of quantum squeezing, i.e., noise reduction below that of zero-point fluctuations in the flexural displacement mode.

4.2.2.2.8 Nanomechanical Laser

This device concept was proposed by Bargatin and Roukes [188]. The fundamental idea is to engineer a laser-like device in which the resonator is realized by a nanomechanical beam, whose tip has been functionalized with a ferromagnetic material, and whose vibration interacts with an adjacent "active" medium containing nuclear spins biased by an external magnetic field, B_0. With the appropriate geometrical configuration, see Fig. 4-8,

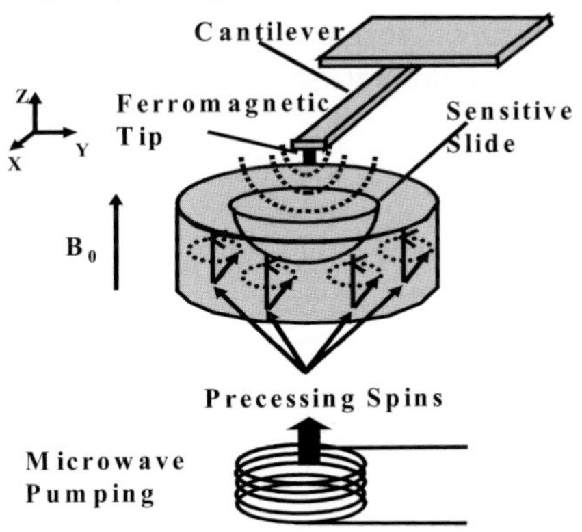

Figure 4-8. Sketch of mechanical laser. (*After* [188].)

vibration of the nanomechanical beam causes superposition of the field produced by its ferromagnetic tip with the external magnetic field.
This results in a modulation of the magnetic field perceived by the nuclear spins and, as a consequence, can stimulate transitions in the Larmor frequency of nuclear spins (Zeeman effect). In turn, a dipolar interaction couples the rotating transverse component of the nuclear magnetization of the nuclear spins with the ferromagnetic tip, resulting in a force that drives the beam oscillations. This process, under resonance between Larmor frequency and beam vibration, leads to self-sustained ocillations, i.e., to laser behavior. The proposed device was called "cantilaser." Typical parameters are as follows: Fundamental frequency of beam, 20 MHz, effective spring constant, 0.1 N/m, quality factor, 10^5, transverse magnetic field gradient due to ferromagnetic tip, $10^6 \mathrm{T/m}$, transverse relaxation time of nuclear spins, $50 \mu\mathrm{s}$, nuclear gyromagnetic ratio, $2\pi \times 10 \mathrm{MHz/T}$, external magnetic field, 2 Tesla.

4.2.2.2.9 Quantum Entanglement Generation

As discussed in Chapter 3, quantum entanglement is a fundamental ingredient for effecting quantum information processing. Most schemes for quantum entanglement, however, were demonstrated in the context of optical experiments, where the object of entanglement was photon polarization. While the realm of implementation of NanoMEMS SoCs includes variants that exploit optical signal processing, i.e., the processing and manipulation of photons, electrons and, thus, electronic signal processing in solid-state systems remain an important paradigm. It is not surprising, therefore, that a number of efforts have been aimed at finding ways to achieve the electron pair entanglement and transport over long distances. The superconductor-carbon nanotube junction, proposed by Bena, Vishveshwara, Balents, and Fisher [189] is a clever idea along these lines, see Fig. 4-9.

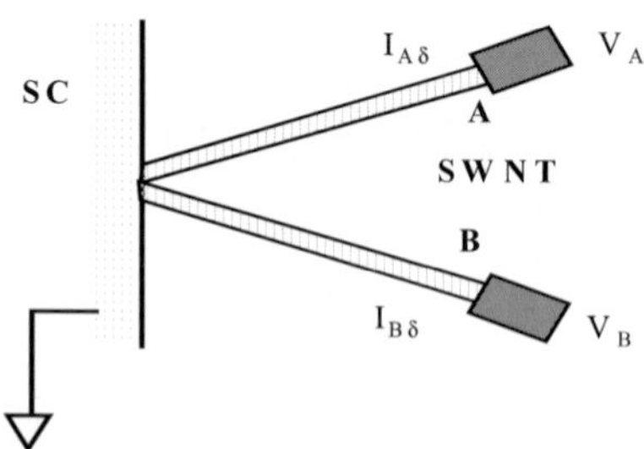

Figure 4-9. Quantum entanglement junction. A setup of two nanotubes A and B contacting a superconductor. Voltage drops V_A and V_B may be preferentially applied across tubes A and B respectively, and currents through each of them may be measured. [189].

The concept consists in exploiting the inherent entanglement of superconducting Cooper pair together with electron-electron interactions in one dimension to enable the sequential injection of entangled pairs from a superconductor into two nanotubes located next to each other at a distance well below the Cooper pair coherence length. The key to the Cooper pair injection and separation into entangled electrons relies on the Lüttinger liquid behavior exhibited by CNTs characterized by an interaction factor *g and* subband spacing ε_0. In particular [189], the tunneling rate, $\Gamma_{AA} \sim (eV/h)(kT/\varepsilon_0)^{\frac{1}{g}-1}$, at which Cooper pairs tunnel from the superconductor into the end of a CNT, being proportional to $eV\rho_{2e}$, turns out to be much smaller than the tunneling rate $\Gamma_{AB} \sim (eV/h)(kT/\varepsilon_0)^{\frac{1}{2}\left(\frac{1}{g}-1\right)}$, at which split entangled pairs are injected into both CNTs. This difference, is rooted in the fact that Lüttinger liquid behavior, manifested as the coherent arrangement of all electrons in the CNT bulk, causes the single-electron tunneling density of states, $\rho_e(E) \sim \varepsilon_0^{-1}(E/\varepsilon_0)^{\frac{1}{4}\left(\frac{1}{g}-1\right)}$ to dominate the Cooper pair tunneling density of states, $\rho_{2e}(E) \sim \varepsilon_0^{-1}(E/\varepsilon_0)^{\frac{1}{g}}$. With $\Gamma_{AA} << \Gamma_{AB}$, virtually all the charge tunneling that occurs involves split entangled pairs.

Once split, the entangled electrons may propagate for long distances due to the ballistic property that characterizes transport in CNTs of Fermi velocity v_F and length L at low temperatures $T < T_\phi = \hbar v_F / k_B L$ at which loss of coherence due to thermal effects are nonexistent.

4.3.1 Quantum Computing Paradigms

As indicated in Chapter 2, the fundamental building block on which quantum information processing systems are based is the qubit, a two-state quantum system. Qubits may take on many physical forms, however, to be useful in realizing real, practical, systems, they must be endowed with three key properties [190]: 1) They must be decoupled from the environment to avoid disturbances which may deviate their time evolution from unitarity; 2) They must be able to respond, in a controlled fashion, to purposeful manipulation, in order to enable the formation of quantum logic gates and entangled states, which rely on such interactions; 3) They must withstand the momentary, but strong, coupling to the environment introduced by a measuring device. In this section, we present the principles of various qubit implementations, in particular, ion-trap-, nuclear-magnetic resonance-, solid-state-, and superconducting-based qubits.

4.3.1.1 The Ion-Trap Qubit

The ion-trap qubit was proposed by Cirac and Zoller [191]. It is embodied by atomic ions confined by an electrode structure designed in such a way that a 3-dimensional harmonic potential well (trap) is produced [190]. Cooling the ions lowers their energy and, were it not because of Coulomb's force of repulsion, which maintains them apart, they would descend to, and meet at, the bottom of the well. Instead, the collective state of the ions is the result of a balance between the potential well energy profile and the forces of repulsion between ions, which manifests in their assuming a linear array, see Figure 4-10.

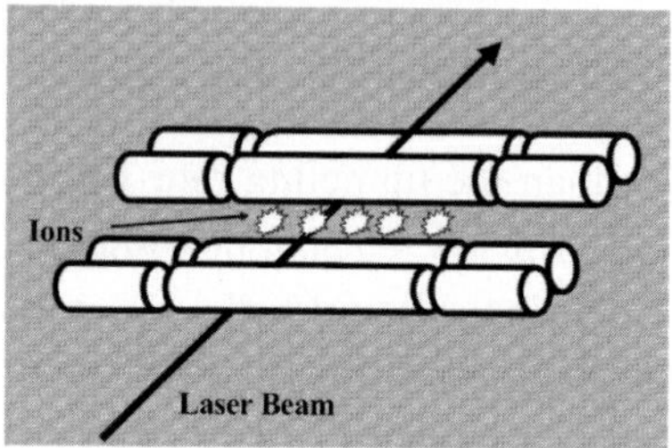

Figure 4-10. Sketch of ion trap qubit. The electrodes create a 3-D harmonic potential well that confines the ions.

The ion trap *simultaneously* implements two types of qubit, Fig. 4-11. In one

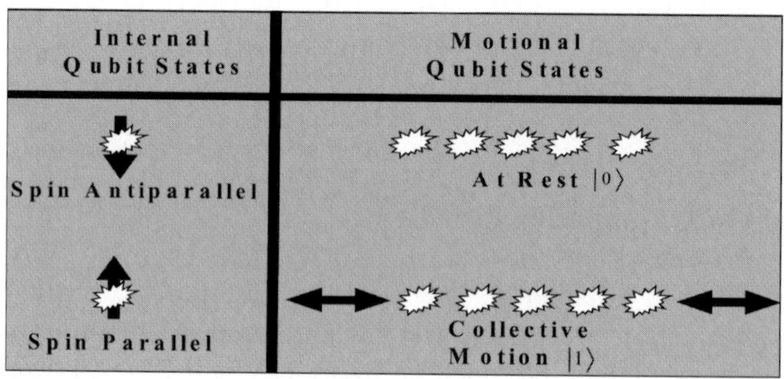

Figure 4-11. Qubit realizations with ion trap.

instance, the two states of the qubit are embodied by the direction of the ion's magnetic moment, which is parallel or antiparallel to an externally applied magnetic field. In the second instance, the collective motion of an array of ions forms the qubit. In particular, when expressed in terms of normal modes, the two states of the motional qubit are the one in which the ions move simultaneously in the same direction, common mode (CM) and the one in which adjacent ions move in opposite directions, stretched mode.

In the motional case, the qubit is not associated with any individual ion, but rather, with the array as a whole. Since the ion trap produces two qubits, a controlled interaction between them allows the realization of quantum gates.

In the case of the spin-orientation qubit, the internal spin state of the ion may be set into the "down" ($|\downarrow\rangle$) or "up" ($|\uparrow\rangle$) states, by application of a uniform magnetic field. Alternatively, it may be prepared into superposition states $\alpha|\downarrow\rangle+\beta|\uparrow\rangle$ by varying the time duration of applied RF fields.

Further functionality is obtained out of the ion-trap system by coupling its spin-orientation qubit to its motional qubit. In particular, superposition of a spatially non-uniform magnetic field along the motional qubit, for instance, of magnitude $+\Delta B$ at the ion's left most position and $-\Delta B$ at its right-most position, causes the ion to experience a field of amplitude ΔB and frequency equal to the motional oscillation frequency. Under these circumstances, an exchange of energy between the spin and the motional states, $|\uparrow\rangle|0\rangle \rightarrow |\downarrow\rangle|1\rangle$, ensues if the magnetic field frequency coincides with the energy difference between the two spin states. More generally, if the spin qubit is in a superposition state, e.g., $\alpha|\downarrow\rangle+\beta|\uparrow\rangle$ then, consistent with conservation of energy, the energy exchange produces the transition $(\alpha|\downarrow\rangle+\beta|\uparrow\rangle)|0\rangle \rightarrow |\downarrow\rangle(\alpha|0\rangle+\beta|1\rangle)$. As depicted in Fig. 4-11,

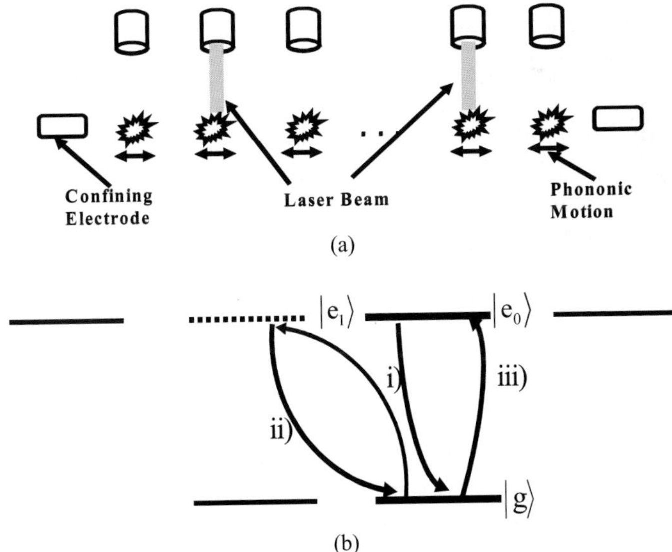

Figure 4-12. (a) Cirac-Zoller ion-trap qubit. (b) Qubit states $|g\rangle$ and $|e_0\rangle$, are separated by an energy $\hbar\omega_0$.

the interaction may be particularized, to the state of one of the ions in the motional qubit, by causing the magnetic field gradient to exist on it. This is accomplished by focusing a laser beam on the ion in question, see Figure 4-12.

Analytically, the inner workings of the ion-trap qubit were described in detail by Cirac and Zoller [191] as follows. The two states of a particular ion, namely, its ground and excited states, are denoted by $|g\rangle_n \equiv |0\rangle_n$ and $|e\rangle_n \equiv |1\rangle_n$, respectively. The 3-dimensional motion confinement of the ions is described by an anisotropic harmonic potential characterized by frequencies $v_x << v_y, v_z$. The typical energy level scheme contemplated for the ion trap is shown in Fig. 4-12(b). When the extent of the ion's motion is much less than the inverse wavevector of the laser field, the so-called Lamb-Dicke limit (LDL), the oscillations of the ground state become normal modes. Under these circumstances, a laser beam with frequency $\omega_L = \omega_0 - \nu_x$, or detuning equal to minus the CM mode frequency, $\delta_n = -v_x$, will excite the common mode exclusively. This is the situation in which transitions $|\downarrow\rangle \rightarrow |\uparrow\rangle$ lead to motional mode (phonon number) transitions $|n\rangle \rightarrow |n-1\rangle$. On the other hand, if $\omega_L = \omega_0 + \nu_x$, then the transition $|\downarrow\rangle \rightarrow |\uparrow\rangle$ leads to $|n\rangle \rightarrow |n+1\rangle$ transitions. Finally, when $\omega_L = \omega_0$ the induced transitions $|\downarrow\rangle \rightarrow |\uparrow\rangle$ leave $|n\rangle$ unchanged. Thus, the relationship between laser detuning, δ and motional frequency, and the fact that the frequencies of the different normal modes are well separated in the excitation spectrum, allows the control of interaction between ions via the CM motion and, in fact constitutes the coupling of two qubits which is necessary to produce quantum gates.

After the quantum qubits are manipulated to effect a quantum computation, the result must be read. In the case of the ion trap this is accomplished by measuring the spin-dependent scattered light when a laser beam impinges upon an ion. Exploiting the fact that scattering is substantially greater for the $|\downarrow\rangle$ spin than for the $|\uparrow\rangle$ spin, the state of the spin is inferred.

The manipulation of the state of an N-ion-trap qubit by a laser beam is driven by the interaction between an ion and the electric field of the laser. Starting with the Hamiltonian for the *n-th* ion, H_0, in the ground state and in the absence of any laser field, and choosing the laser frequency as above, i.e., $\delta_n = -v_x$, and the ion position to coincide with a node of the laser standing wave, the system is described by,

$$H_{n,q} = \frac{\eta}{\sqrt{N}} \frac{\Omega}{2} \left[\left| e_q \right\rangle_n \left\langle g \right| a e^{-i\varphi} + \left| g \right\rangle_n \left\langle e_q \right| a^+ e^{i\varphi} \right], \tag{6}$$

where a and a^+ are the creation an annihilation operators of CM phonons, respectively, Ω is the Rabi frequency, ϕ is the phase of the laser field at the mean position of the ion, $q = 0,\ 1$ levels involved in the energy transition excited by the laser, and $\eta = \sqrt{\hbar k_\theta^2 / 2Mv_x} << 1$ is the LDL parameter, with $k_\theta = k cos\theta$, k the laser wavevector, θ the angle between the direction of propagation of the laser and the x-axis of motion of the qubit, and M the ion mass. The Rabi frequency, $\Omega = -E_0 \left\langle \uparrow \left| \vec{d} \cdot \hat{\varepsilon}_L \right| \downarrow \right\rangle / 2\hbar$, characterizes the transition frequency between the ground and metastable states produced by a laser with electric field amplitude E_0 and polarization vector $\hat{\varepsilon}_L$ in an ion of electric dipole operator $\vec{d}$.

The evolution of the system upon being impinged by a laser beam pulse of time duration $t = k\pi / \left(\Omega\eta / \sqrt{N} \right)$ on the *n-th* ion is described by the unitary operator,

$$\hat{U}_n^{k,q}(\varphi) = \exp\left[-ik \frac{\pi}{2} \left(\left| e_q \right\rangle \left\langle g \right| a e^{-i\varphi} + \left| g \right\rangle_n \left\langle e_q \right| a^+ e^{i\varphi} \right) \right]. \tag{7}$$

Application of this unitary operator on the various states of the *n-th* qubit yields the results of Table 4-1. $|0\rangle$ and $|1\rangle$ represent the population of the CM mode with zero and one phonon, respectively.

Table 4-1. Effect of Ion-Trap Unitary Operator on State Evolution

Operator	Initial State	Final State
$\hat{U}_n^{k,q}$	$\|g\rangle_n \|0\rangle$	$\|g\rangle_n \|0\rangle$
$\hat{U}_n^{k,q}$	$\|g\rangle_n \|1\rangle$	$cos(k\pi/2) \|g_n\rangle \|1\rangle - ie^{i\varphi} sin(k\pi/2) \|e_q\rangle_n \|0\rangle$
$\hat{U}_n^{k,q}$	$\|e\rangle_n \|0\rangle$	$cos(k\pi/2) \|e_q\rangle_n \|0\rangle - ie^{-i\varphi} sin(k\pi/2) \|g\rangle_n \|1\rangle$

The above interaction is amenable to the implementation of a two-bit gate. In particular, Cirac and Zoller [191] have shown that this is accomplished by following three steps: 1) Apply a π laser pulse with polarization $q = 0$ and phase $\phi = 0$ to the *m-th* ion to create the evolution $\hat{U}_m^{1,0} \equiv \hat{U}_m^{1,0}(0)$; 2) Turn on the laser directed to the *n-th* ion for a time duration 2π and polarization and phase $q=1$ and phase $\phi=0$, respectively. This creates the evolution operator $\hat{U}_{n}^{2,1}$, which exclusively changes the sign of the sate $|g\rangle_n|1\rangle$ via a rotation through the state $|e_1\rangle_n|0\rangle$; 3) Apply again a π laser pulse with polarization $q=0$ and phase $\phi=0$ to the *m-th* ion to create the evolution $\hat{U}_{m}^{1,0} \equiv \hat{U}_{m}^{1,0}(0)$. Since these operators act on non-interacting ions, the overall effect is given by the product $\hat{U}_{m,n} \equiv \hat{U}_{m}^{1,0}\hat{U}_{n}^{2,1}\hat{U}_{m}^{1,0}$ in Eq. (8) below. Comparison of the first and last columns reveals that the effect of the composite operation is to change the sign of the state only when both ions are initially excited, thus, Eq. (8) embodies a C-NOT gate.

$$
\begin{array}{cccc}
 & \hat{U}_{m}^{1,0} & \hat{U}_{n}^{2,1} & \hat{U}_{m}^{1,0} \\
|g\rangle_m|g\rangle_n|0\rangle \rightarrow & |g\rangle_m|g\rangle_n|0\rangle \rightarrow & |g\rangle_m|g\rangle_n|0\rangle \rightarrow & |g\rangle_m|g\rangle_n|0\rangle \\
|g\rangle_m|e_0\rangle|0\rangle \rightarrow & |g\rangle_m|e_0\rangle|0\rangle \rightarrow & |g\rangle_m|e_0\rangle|0\rangle \rightarrow & |g\rangle_m|e_0\rangle|0\rangle \\
|e_0\rangle_m|g\rangle_n|0\rangle \rightarrow & -i|g\rangle_m|g\rangle_n|1\rangle \rightarrow & i|g\rangle_m|g\rangle_n|1\rangle \rightarrow & |e_0\rangle_m|g\rangle_n|0\rangle \\
|e_0\rangle_m|e_0\rangle_n|0\rangle \rightarrow & -i|g\rangle_m|e_0\rangle_n|1\rangle \rightarrow & -i|g\rangle_m|e_0\rangle_n|1\rangle \rightarrow & -|e_0\rangle_m|e_0\rangle_n|0\rangle
\end{array} . \quad (8)
$$

Many successful implementations of ion-trap qubits have been experimentally demonstrated [192]. Key to these experimental demonstrations are techniques to address a variety of issues, most notably: 1) Mitigating the decoherence of the ion trap, which is due to the spontaneous decay of the internal atomic states and the motion damping; 2) Suppressing spontaneous emission; 3) Obtaining highly efficient read-out schemes. A thorough discussion of problems and solutions regarding ion-trap qubits is given by Wineland et al [192].

4.3.1.2 The Nuclear Magnetic Resonance (NMR) Qubit

As is well known, some atoms exhibit an intrinsic nuclear magnetic moment $\vec{\mu}$ and an angular momentum $\hbar\vec{I}$, and these are related through the gyromagnetic ratio γ by [28],

$$\vec{\mu} = \gamma\hbar\vec{I} . \quad (9)$$

Since the angular momentum is quantized [60], with values $m_I = I, I-1,\ldots,-I$, a nucleus with an intrinsic angular momentum of half a unit, i.e., $I = 1/2$, will have the allowed values of $m_I = \pm 1/2$. Thus, in the presence of a magnetic field $\vec{B} = B_0\hat{z}$, the energy of interaction between the magnetic moment and the field,

$$H_I = -\vec{\mu} \cdot \vec{B} = -\gamma\hbar B_0 I_z , \qquad (10)$$

will split into two energy levels, see Fig. 4-13 at the top of next page.

These two energy levels in a non-zero field embody a two-state quantum system that can be used as a qubit. The controlled manipulation of these qubits to effect quantum computations is the goal of NMR-based quantum computing (QC). The origins, development, progress and status of NMR-based QC has been addressed recently in extensive review articles by Laflamme, Knill, Cory *et al.* [193], and by Vandersypen and Chuang [194]. Our presentation, therefore, will follow these closely.

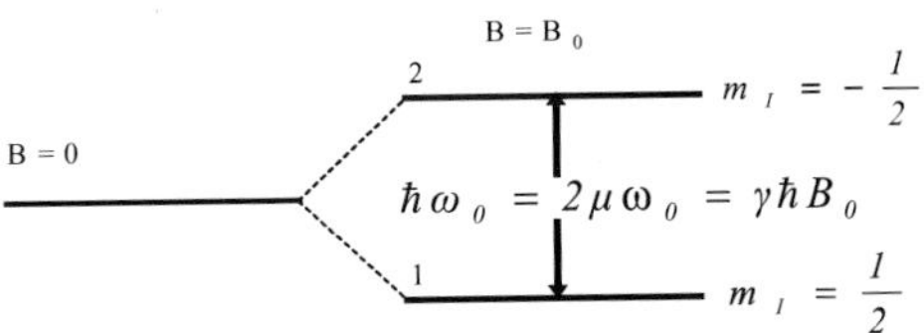

Figure 4-13. Energy level splitting when a nucleus of intrinsic angular momentum $I = 1/2$ is exposed to a constant magnetic field B_0.

In practice, limits germane to currently available techniques preclude detecting the energy absorbed by a single nucleus. Therefore, a substance containing a multitude of nuclei, whose contributions add, must be employed [193]. The system of choice for NMR-based QC consists of the very large number of nuclei belonging to atoms forming a molecule in a liquid, so-called liquid-state NMR. Fig. 4-14 depicts a typical molecule used to form

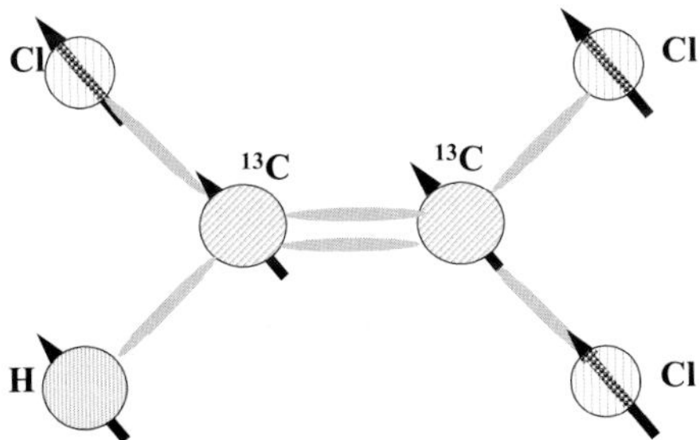

Figure 4-14. Trichloroethylene molecule for liquid-state NMR-based QC. The proton (H), and the two carbons (^{13}C) are employed to realize qubits. The ^{13}C nucleus has spin ½. [193].

qubits is the trichloroethylene (TCE) molecule, which contains a hydrogen nucleus possessing a strong magnetic moment. As a result, when the molecule is exposed to a constant strong magnetic field, B, each hydrogen's spin orients itself in the direction of the field. If, in addition, an RF field is applied in a pulsed fashion, the spins are made to tip off-axis, while precessing about the direction of the constant field. The precession frequency is the so-called Larmor frequency and is given by $\omega = \mu B$. For the hydrogen atom (proton), the magnetic moment is 42.7MHz/T and, at a typical field of B=11.7T, its precession frequency is 500MHz. Sample examination is accomplished placing a coil around it, tuned to the precession frequency, which picks up the oscillating currents induced as a consequence of the magnetic field produced by the precessing protons. The device that applies the static magnetic field and the RF control pulses, and then detects the magnetic induction is called an NMR spectrometer, see Fig. 4-15.

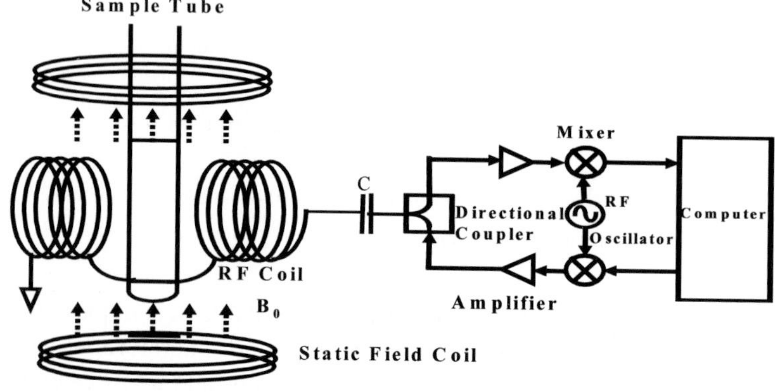

Figure 4-15. Sketch of experimental NMR spectrometer. (*After* [195].)

NMR phenomena, which were first observed in 1946 [196], [197], became the basis for a multitude of analytical studies of materials, in particular, the determination of molecular structures [198], and magnetic resonance imaging [199]. In these contexts, the technology of NMR *spectroscopy* is rather mature.

The application of NMR to QC was advanced Cory, Fahmy and Havel [200], and Gershenfeld and Chuang [201] in 1997. To overcome the difficulty in detecting the spin of individual, adoption was made of qubits implemented in the liquid state, where additive effects could be exploited to yield a reasonably large signal amplitude. Also adopted were methods to discern the fraction of nuclear spins pointing in the external field direction, despite the effects of temperature-induced random spin orientation. The two-state quantum system was realized by choosing molecules possessing spin-1/2 nuclei, which in the presence of the external magnetic field adopts two states, namely, a low energy state denoted by $|0\rangle$, and a higher energy state denoted by $|1\rangle$.

Analytically, an NMR-based QC system is described in terms of two Hamiltonians, namely, the system Hamiltonian, which captures the energy of single and coupled spins in the presence of a magnetic field, and the control Hamiltonian, which captures the effects of applied RF pulses controlling the operations with qubits.

The system Hamiltonian for single spins is given by,

$$H_0 = -\hbar\gamma B_0 I_z = -\hbar\omega_0 I_z = \begin{bmatrix} -\hbar\omega_0/2 & 0 \\ 0 & \hbar\omega_0/2 \end{bmatrix}, \tag{11}$$

where I_z is the z-component of the angular momentum $I = \hat{x}I_x + \hat{y}I_y + \hat{z}I_z$. In general, the three components of the angular momentum are related to the Pauli spin matrices as follows [60],

$$\sigma_x = 2I_x,\ \sigma_y = 2I_y;\ \sigma_z = 2I_z, \tag{12}$$

where,

$$\sigma_x \equiv \begin{bmatrix} 0 & 1 \\ 1 & 0 \end{bmatrix},\ \sigma_y \equiv \begin{bmatrix} 0 & -i \\ i & 0 \end{bmatrix};\ \sigma_z \equiv \begin{bmatrix} 1 & 0 \\ 0 & -1 \end{bmatrix}. \tag{13}$$

H_0 embodies the time evolution given by the $U = e^{-iH_0 t/\hbar}$, which represents the precession of the overall state vector (the so-called Bloch vector) with respect to the axis $\vec{B}$, defined by the static magnetic field, see Fig. 4-16 [194].

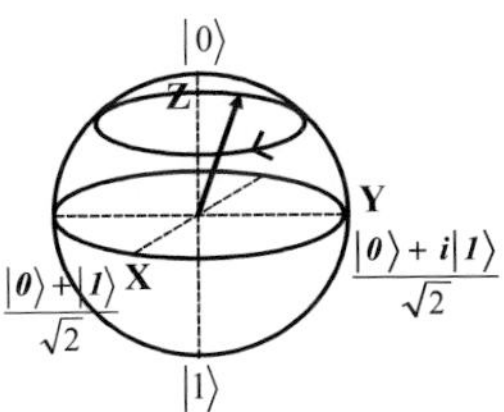

Figure 4-16. Precession of a spin-1/2 about the axis of a static magnetic field. (*After* [194].)

Vandersypen and Chuang [194] indicate that in the most general case, the system Hamiltonian for a molecule possessing N isolated nuclei is given by,

$$H_0 = -\sum_{i=1}^{N} \hbar(1-\bar{\sigma}_i)\gamma_i B_0 I_z^i = -\sum_{i=1}^{N} \hbar\omega_0^i I_z^i, \tag{14}$$

where i labels the nuclei, and $\bar{\sigma}_i$ denotes the so-called chemical shifts, which characterize the fact that distinctly different precession frequencies are exhibited by identical atomic species within a given molecule, when the shielding environment produced by their surrounding electrons results in a different magnetic field, B_0. They also point out [194] that typical chemical shifts range in the order of a few kilohertz, compared to the precession frequencies, which range in the MHz.

In addition to isolated spin nuclei, liquid-state NMR includes the presence of coupled spins. These are characterized by either a *direct* or an *indirect* coupling mechanism. The direct coupling is of the magnetic dipole-dipole nature, similar to the interaction between two adjacent bar magnets and, for nuclei i and j, separated by a distance r_{ij}, is given by [194],

$$H_D = \sum_{i<j} \frac{\mu_0 \gamma_i \gamma_j \hbar}{4\pi |\vec{r}_{ij}|^3} \left[\vec{I}^i \cdot \vec{I}^j - \frac{3}{|\vec{r}_{ij}|^2} \left(\vec{I}^i \cdot \vec{r}_{ij} \right) \left(\vec{I}^j \cdot \vec{r}_{ij} \right) \right], \tag{15}$$

where μ_0 is the free space magnetic permeability, and $\vec{\mathrm{I}}^{\mathrm{i}}$ is the magnetic moment vector of spin i. Under certain conditions, Eq. (15) may be simplified. For instance, for large precessing frequencies it reduces to [194],

$$H_D = \sum_{i<j} \frac{\mu_0 \gamma_i \gamma_j \hbar}{8\pi |\vec{r}_{ij}|^3} \left(1 - 3\cos^2 \theta_{ij} \right) \left[3\vec{I}^i \cdot \vec{I}^j - \left(\vec{I}^i \cdot \vec{I}_{\mathrm{j}} \right) \right], \tag{16}$$

where θ_{ij} is the angle between B_0 and $\vec{r}_{ij}$, whereas if $\left| \omega_0^{\mathrm{i}} - \omega_0^{\mathrm{j}} \right|$ is much greater than the coupling strength it reduces to [191],

$$H_D = \sum_{i<j} \frac{\mu_0 \gamma_i \gamma_j \hbar}{4\pi |\vec{r}_{ij}|^3} \left(1 - 3\cos^2 \theta_{ij} \right) \mathrm{I}_{\mathrm{z}}^{\mathrm{i}} \mathrm{I}_{\mathrm{z}}^{\mathrm{j}} . \tag{17}$$

The indirect coupling is characterized by a strength *J*, which captures the overlap of electronic wavefunctions between two atomic nuclei, and has values ranging from several Hz, for three- to four-bond couplings, to several KHz for one-bond coupling. The indirect coupling Hamiltonian takes the form [194],

$$H_J = \hbar \sum_{i<j} 2\pi\pi_{ij} \vec{I}^i \vec{I}^j = \hbar \sum_{i<j} J_{ij} \left(I_x^i I_x^j + I_y^i I_y^j + I_z^i I_z^j \right), \tag{18}$$

where J_{ij} characterizes the coupling between spins *i* and *j*. Simplification of this expression is also possible in certain circumstances, in particular, when

$\left|\omega_i - \omega_j\right| >> 2\pi\left|J_{ij}\right|$, which may be obtained when dealing with heteronuclear spins or with small homonuclear molecules, it reduces to [194],

$$H_J = \hbar\sum_{i<j}^{N} 2\pi J_{ij} I_z^i I_z^j \,. \tag{19}$$

Eq. (19) captures the circumstance that, in addition to a constant externally applied magnetic field, $\vec{B}$, the actual field at a given spin location includes a static field along $\pm\hat{z}$, which is elicited by spins in its neighborhood. The consequence of this additional field is to shift the spin's energy levels and manifests as a change in the Larmor frequency. For instance, a neighboring spin *j* in state $\left|0\right\rangle$ will shift the frequency of spin *i* by $-J_{ij}/2$, whereas if spin *j* is in state $\left|1\right\rangle$, it will shift the frequency of spin *i* by J_{ij} $+J_{ij}/2$. In general, it turns out that, when in the presence of neighboring spins, the spectrum of a given spin would show, instead of a *single* line at its Larmor frequency, two lines for every neighboring spin, the lines being separated by the coupling strength J_{ij} and located equidistantly above and below the Larmor frequency.

In the majority of NMR-based QC experiments, the system Hamiltonian realized is described by the simplified Hamiltonians [194], i.e.,

$$H_{sys} = -\sum_i \hbar\omega_0^i I_z^i + \hbar\sum_{i<j} 2\pi J_{ij} I_z^i I_z^j \,, \tag{20}$$

where the first term arises from the energy of isolated spins, and the second from the energy of interacting (coupled) spins.

To effect the manipulation of qubits in NMR-based QC [194], it is necessary to apply a magnetic field that will rotate the state of the spin-1/2 nuclei, see Fig. 4-16. This is accomplished by adding to the static $\hat{z}$-directed magnetic field, B_0, a time-varying (RF) electromagnetic field oriented in the $\hat{x}-\hat{y}$ plane, of a frequency ω_{RF} close to the spin precession frequency ω_0. This RF field gives rise to the *control* Hamiltonian which, for a single spin, is given by [194],

$$H_{RF} = -\hbar\gamma B_1\left[cos(\omega_{RF}t+\varphi)I_x + sin(\omega_{RF}t+\phi)I_y\right], \tag{21}$$

where B_1 is the applied RF field amplitude and ϕ its phase. For liquid-state NMR, $\gamma B_1 \approx 50\text{KHz} \equiv \omega_1$. In the presence of N spins, the total control Hamiltonian is the sum of the terms such as Eq. (21) of each spin. The implementation of quantum gates in NMR-based QC exploits the ability to induce a certain time evolution of a spin state by the fine perturbation that varying the amplitude, frequency, and phase of the control Hamiltonian affords.

The analysis of spin rotations is facilitated by describing the motion with respect to the so-called rotating frame [193], [194]. This is a coordinate system that rotates with respect to the $\hat{z}$ axis at a frequency ω_{RF}. A given state in the rotating frame $|\psi\rangle^{rot}$ and the corresponding state $|\psi\rangle$ in the laboratory (non-rotating) frame are related by [191],

$$|\psi\rangle^{rot} = \exp(-i\omega_{RF} t I_z)|\psi\rangle. \tag{22}$$

It can be shown by substitution of (22) into Schödinger's equation, that in the rotating frame and in the presence of many, e.g., *r*, applied RF fields, the system and control Hamiltonians adopt the forms [194],

$$H_{sys} = \hbar \sum_{i<j} 2\pi J_{ij} I_z^i I_z^j, \tag{23}$$

and

$$H_{control} = \sum_{i,r} -\hbar\omega_1^r \left[\cos\left(\left(\omega_{RF}^r - \omega_0^i\right)t + \phi^r\right) I_x^i + \sin\left(\left(\omega_{RF}^r - \omega_0^i\right)t + \phi^r\right) I_y^i\right]. \tag{24}$$

The effect of the control Hamiltonian is most easily visualized with reference to the Bloch sphere, see Fig. 4-17, whose surface contains the

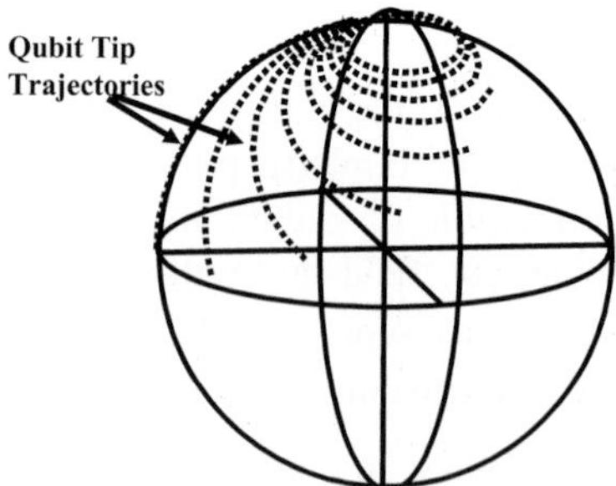

Figure 4-17. Bloch sphere surface: Dashed lines delineate the trajectories of the tip of a qubit as a function of the RF pulse strength and duration. When the RF frequency equals the Larmor frequency, i.e., at resonance, the pulse produces a 90° rotation. As $|\omega_{RF} - \omega_0|$ increases, the rotation decreases, in particular, at large offsets the trajectory remains close to $|0\rangle$. [194].

locus of the tip of a qubit vector as a function of $|\omega_{RF} - \omega_0|$, for a given RF pulse duration and the parameter ω_1.

NMR-based quantum gates are generated by "tuning" the parameters in the control Hamiltonian to achieve a desired qubit rotation. Since any quantum gate may be constructed from single-qubit rotations and the C-NOT gate, the problem of NMR-based quantum computing reduces to determining the control Hamiltonian that will implement these. In this context, we note that the most general qubit rotation is defined by [194],

$$R_{\hat{n}} = exp\left[-\frac{i\theta\hat{n}\cdot\vec{\sigma}}{2}\right], \tag{25}$$

where $\hat{n}$ denotes the 3-dimensional axis of rotation, θ is the angle of rotation, and $\vec{\sigma} = \hat{x}\sigma_x + \hat{y}\sigma_y + \hat{z}\sigma_z$ is a vector of Pauli matrices. Furthermore, it can be shown that any qubit transformation may be implemented as a sequence of rotations about only two axes. In particular, Bloch's theorem stipulates such a transformation as [194],

$$U = e^{i\alpha}R_x(\beta)R_y(\gamma)R_x(\delta). \tag{26}$$

Therefore, in terms of the control Hamiltonian parameters, implementing a single-qubit gate may be accomplished in the rotating frame using RF pulses. Specifically, if an RF field of amplitude ω_1 and frequency is $\omega_{RF} = \omega_0$ is applied to a single spin, this will evolve according to [194],

$$U = \exp\left[i\omega_1\left(\cos\phi I_x + \sin\phi I_y\right)t_{pulse}\right], \tag{27}$$

where the RF pulse duration is given by t_{pulse}. In the context of the Bloch sphere, this transformation would rotate the qubit by an angle $\theta \sim \omega_1 t_{pulse}$, with respect to an axis in the $\hat{x} - \hat{y}$ plane given by the phase ϕ. For instance, the parameters: $\phi = \pi$ and $\omega_1 t_{pulse} = \pi/2$ effect the $R_x(90)$ rotation about $\hat{x}$, whereas doubling the pulse duration implements $R_x(180)$, and changing the phase to $\phi = -\pi/2$ effects the rotation about $\hat{y}$. In general, the phase of the RF pulse determines the nutation axis in the rotating frame, so that to perform $\hat{x}$ and $\hat{y}$ rotations it is not necessary to orient the RF field along these axes; changing the phase suffices. A rotation about the $\hat{z}$ axis in terms of rotations about $\hat{x}$ and $\hat{y}$ is given by [194],

$$U = R_z(\theta) = XR_y(\theta)\overline{X} = YR_x(-\theta)\overline{Y}, \tag{28}$$

where the bar over X and Y denotes a rotation of -90 degrees with respect to X or Y.

The NMR-based implementation of the C-NOT quantum gate involves a series of two-qubit rotations, namely [201]: $|00\rangle \rightarrow |00\rangle$, $|01\rangle \rightarrow |01\rangle$, $|10\rangle \rightarrow |11\rangle$, and $|11\rangle \rightarrow |10\rangle$. Addressing a particular qubit, without affecting the neighboring one, is accomplished by exploiting the fact that different atoms possess different resonance frequencies, ω_0, or that the same type of atoms with a different chemical shift also possess different ω_0. Taking two coupled spin-1/2 atoms with resonance frequencies ω_1 and ω_2, and coupling J_{12}, the C-NOT gate is implemented if applying a narrowband 180-degree pulse at a frequency $\omega_2 + J_{12}/2$, causes spin 2 to be inverted only if spin 1 is in state $|1\rangle$. In this case, spin 1 is the control qubit and spin 2 the target qubit. Pictorially, the C-NOT gate may be visualized following the construction of Steffen, Vandersypen and Chuang [201], see Fig. 4-18. The sequence of rotations is produced as follows: 1) An RF pulse at a frequency ω_2, of a bandwidth such covering the frequency range $\omega_2 \pm J_{12}$, but that does not overlap with ω_1, rotates spin 2 from +Z to –Y; 2) The spin system is allowed to evolve freely for a duration of $1/2J_{12}$ seconds; 3) During the free evolution period, the precession frequency of spin 2 will be shifted by $\pm J_{12}/2$ according to whether spin 1 is in the $|1\rangle$ or $|0\rangle$ state. This will result in the rotation of spin 2 to either +X or –X by the end of this period, depending on the state of spin 1; 4) A 90-degree pulse applied to spin 2 about the –Y axis rotates spin 2 to +Z if spin 1 is in state $|0\rangle$, or to –Z if spin 1 is in state $|1\rangle$.

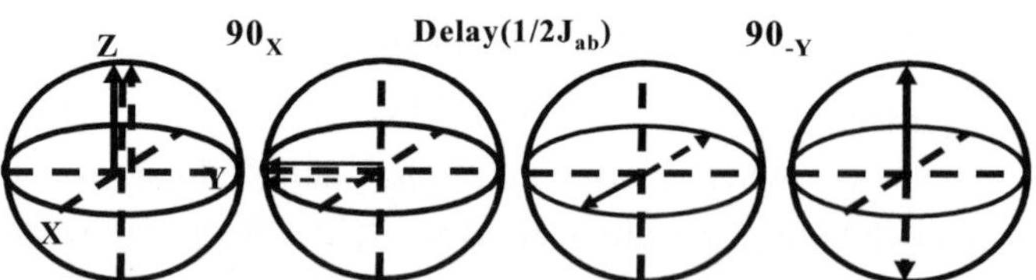

Figure 4-18. Left-to-right: Sequence of qubit rotations for implementing the C-NOT quantum gate in NMR-based QC. The coordinate system rotates around the $\hat{Z}$ axis at a frequency ω_2 when spin 1 is $|0\rangle$ (solid line), and $|1\rangle$ (dashed line). (*After* [201].)

While the maturity of NMR spectroscopy has enabled the successful proof-of-concept implementation of various QC algorithms, the fact that the technique must rely on measuring ensembles of spins to obtain a detectable read-out signal is a limiting aspect of it, since this implies that one must begin with the highly-mixed initial ensemble state; this is the result of there being a very small energy difference between up and down spins at room temperature, manifesting itself as a nearly random equilibrium distribution [193].

A highly-mixed state possesses equally likely spin-up and spin-down states, for example [193],

$$(1-\varepsilon)\boldsymbol{I}/2+\varepsilon|0\rangle\langle 0|, \tag{29}$$

$\varepsilon \sim 10^{-5}$, which is an almost random state with a small excess of the $|0\rangle$ state [193]. This expression for the equilibrium state follows from the density matrix $\boldsymbol{\rho}_{thermal}$ which, being proportional to $e^{-H/kT}$ (where the nuclear spins in a molecule posses the internal Hamiltonian H, T is temperature and k is the Boltzmann constant), admits an expansion [190],

$$e^{-H/kT} \approx e^{-\varepsilon_1\sigma_z^{(1)}/kT} e^{-\varepsilon_2\sigma_z^{(2)}/kT} \ldots, \tag{30}$$

which with,

$$\boldsymbol{e}^{-\varepsilon_1\sigma_z^{(1)}/kT} \approx \boldsymbol{I} - \varepsilon_1\sigma_z^{(1)}/\boldsymbol{kT}\ldots, \tag{31}$$

may be written as,

$$\boldsymbol{e}^{-H/kT} \approx \boldsymbol{I} - \varepsilon_1\sigma_z^{(1)}/\boldsymbol{kT} - \varepsilon_2\sigma_z^{(2)}/\boldsymbol{kT}\ldots, \tag{32}$$

where $\boldsymbol{I}$ is the identity matrix and, for spin *i*, the parameter ε_i represents the energy difference between up and down states. While the desired initial state is a pure one, in which all spins are in the same state, e.g., $|0\rangle$, the actual randomness of the initial ensemble state may be overcome by a technique to transform it into an almost pure state.

An almost pure state is one that produces a signal that is proportional to that of a pure-state signal. It is generated by exploiting three facts [193], namely: 1) That the magnetization is determined by the traceless part of the density matrix; 2) That the completely mixed state $I/2^n$ is preserved under both unitary and non-unitary transformations; and 3) That all scales are relative, in particular, that only the ratio of two magnetizations determines the final answer of a quantum computation, i.e., the deciding factor in a measurement is, not the absolute magnetization, but its relative value compared to the noise [193].

Constructing a pseudo-pure state makes use of the concept of *deviation density matrix*. This is the arbitrary matrix δ for which $\delta - \rho = \lambda I$ for some constant λ. From this definition, and inspection of Eq. (29), it is clear that the matrix $\varepsilon|0\rangle\langle 0|$ is in fact a deviation matrix from the equilibrium state of one nuclear spin. An interesting property of the deviation matrix is that, if $\hat{m}$ is a traceless operator, then [193],

$$\begin{aligned} \mathrm{tr}(\delta\hat{m}) &= \mathrm{tr}((\rho+\lambda I)\hat{m}) \\ &= \mathrm{tr}(\rho\hat{m}) + \mathrm{tr}(\hat{m}). \\ &= \mathrm{tr}(\rho\hat{m}) \end{aligned} \tag{33}$$

Thus, the expectation value (the measurement) of a traceless observable may be obtained either from the density matrix or from the deviation matrix, as prescribed in Eq. (33). A pseudo-pure state, in fact, is defined as one whose equilibrium state has the deviation $\delta = \varepsilon|0\rangle\langle 0|$. Its significance is as follows. If we are interested in the probability of p_1 of measuring state $|1\rangle$, given that the initial state was $|0\rangle$, then this is given by [193],

$$\begin{aligned} p_1 &= \langle 1|U|0\rangle\langle 0|U^+|1\rangle \\ &= \mathrm{tr}(U|0\rangle\langle 0|U^+|1\rangle\langle 1|) \\ &= \mathrm{tr}(U|0\rangle\langle 0|U^+(I-\sigma_z))/2 \\ &= (\mathrm{tr}(U|0\rangle\langle 0|U^+) - \mathrm{tr}(U|0\rangle\langle 0|U^+\sigma_z))/2 \\ &= (1-\mathrm{tr}(U|0\rangle\langle 0|U^+\sigma_z))/2 \end{aligned} \quad , \tag{34}$$

Where U is the total unitary operator associated with a computation. Therefore, Eq. (34) indicates that by measuring the initial and final expectation values of σ_z, $a = tr(\delta\sigma_z) = \varepsilon$, and $a' = tr(\delta'\sigma_z) = \varepsilon tr(U|0\rangle\langle 0|U^+\sigma_z)$, respectively, one can determine p_1. In fact, $p_1 = (1-(a/a'))/2$, independent of the scale ε.

Most importantly, the technique may be extended to the case in which one desires to determine the probability p_1 of measuring the state $|1\rangle_1$, in the case in which this state refers to the first qubic resulting from applying a quantum gate to an initial state $|000...000\rangle$ [193]. The result is the same, namely, $p_1 = (1-(a/a'))/2$, except that now the deviation in question takes the form $\delta = \varepsilon|000...000\rangle\langle 000...000|$.

In general, if a state has a deviation proportional to a pure state $|\psi\rangle\langle\psi|$, in particular, $\delta = \varepsilon|\psi\rangle\langle\psi|$, it is called a pseudo-pure state. Physically, Cory *et al.* [200] stated that the justification for constructing a pseudo-pure state derives from the fact that the spins in the different molecules of a liquid are virtually independent of one another and that, as a result, they may be construed as a large number of copies of a single type of molecule, thus permitting the liquid to be approximated by a Gibbs ensemble. Because of this, instead of dealing with a density matrix of size 2^N, which is the total number of molecules, one can deal with a reduced density matrix of size 2^n, where *n* is the number of spins in a single molecule. Analytically, instead of the density matrix [200],

$$\Psi = \int_{\{\psi\}} p(\psi)|\psi\rangle\langle\psi| d\psi, \tag{35}$$

where $p(\psi)$ is the probability density of the pure state described by the spinor ψ and $\{\psi\}$ denotes the set of all unit norm spinors, one uses the approximation [200],

$$\Psi = \frac{(1-\alpha)I + 2\alpha|\psi\rangle\langle\psi|}{(1-\alpha)2^n + 2\alpha} \quad (-1 \leq \alpha \leq 1), \tag{36}$$

where $|\psi\rangle$ is a unit spinor. Thus, since the ensemble average of an observable O is obtained by taking the trace of its product by the density matrix, $\mathrm{tr}(O\Psi)$, a simplification is obtained from using the pseudo-state, since the ensemble average is now given by, $\mathrm{tr}(O\Psi) \propto (1-\alpha)\mathrm{tr}(O) + 2\alpha\langle\psi|O|\psi\rangle$, where $\mathrm{tr}(O)$ is known. While the pseudo-pure state continues to be made up of a statistical mixture of molecules, since by Eq. (36), each spinor determines a unique psudo-pure density matrix, and each pseudo-pure density matrix determines a spinor that is unique to within an overall phase factor (assuming the polarization is α known), each addition of the magnetizations of all the molecules reveals the predominance of one particular state present, in effect capturing each molecule's state for the final spectrum without the necessity of wavefunction collapse [200]. The price paid as a result of using pseudo-pure states is the loss of a factor of the order of one million in the effective number of molecules per state because the net polarization of spins is only about one part in one million. Herein lies one of the main limitations of NMR-based QC [193], [200]: The fact that the pseudo-pure state signal decreases exponentially with the number of qubits prepared, while the noise level remains constant, precludes the methods for extracting pseudo-pure states from working for more than about 10 nuclear spins.

Thus, the use of pseudo-pure states enables one to obtain a result despite the highly random nature of the initial state. The question then becomes, how does one transform an initial random state into a pseudo-pure state with deviation $|000...000\rangle\langle 000...000|$? A technique, among various, that is employed applies magnetic field gradient to the sample in order to make the frequency of the precessing spins position-dependent and, thus, make it possible to distinguish different parts of the sample. In particular, the gradient field induces a position-dependent phase change along the sample. This is the basis of NMR imaging [193].

Another issue that derives from the ensemble nature of the sample, is that care must be taken to reduce unintended coupling between qubits [193]. The established technique to accomplish this is called "refocusing" [193], [194]. The fundamental idea is to apply a pulse at the midpoint of the evolution period to a given spin, of such a phase (typically $180°$) as to undue the evolution it has experienced over the time period due to the influence of the undesired coupling [193].

One common issue with QC is the effect of decoherence. In the case of NMR-based QC decoherence is characterized in terms of two parameters, namely, the energy relaxation rate, T_1, and the phase randomization rate, T_2 [194]. T_1 captures the energy lost by precessing spins to various mechanisms such as couplings to other spins, and to phonons and paramagnetic ions, and chemical reactions such as ions exchanges with the solvent. This source of decoherence may, by properly choosing the molecules and liquid samples, be extended to tens of seconds. T_2 captures energy losses due to short- and long-range spin-spin couplings, the effects of fluctuating magnetic fields due to the spatial anisotropy of the chemical shifts, local paramagnetic ions, or unstable laboratory fields. These factors, by properly choosing the quality of the samples and laboratory equipment allow a decoherence time of one second or more for molecules in solution [194].

4.3.1.3 The Semiconductor Solid-State Qubit

Given the predominance of solid state silicon electronics technology, there is a strong motivation to discover and develop paradigms for quantum computing that exploit qubits embedded in silicon wafers. An early example of this is the scheme for a silicon-based nuclear spin quantum computer introduced by Kane [202], see Fig. 4-19. In this section this example is reexamined.

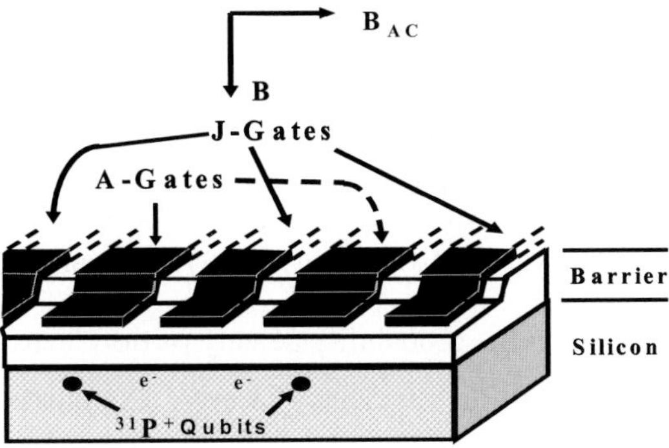

Figure 4-19. Sketch of nuclear spin QC concept. Illustrated are two cells in a one-dimensional array containing $^{31}P^{+}$ donors and electrons in a Si host wafer, separated by a barrier from metal gates on the surface. $B_{ac} \sim 10^{-3}$ Tesla, and $B \sim 2$ Tesla. (*After* [202].)

In this scheme the qubits are embodied in the nuclear spins of donor atoms located underneath biasing metallic gates in doped silicon structures, and the coupling between qubits is enabled by the hyperfine interaction, which couples electron and nuclear spins. In particular, with the wave function of the donor electron being concentrated at the nucleus, a large hyperfine energy, and thus coupling, between electron and nuclear spins is guaranteed which, in turn, may be communicated to adjacent qubits by the extension/overlap of the electron wave functions of the corresponding donor electrons. Modulation of the coupling between electronic wave functions, and thus between qubits, is facilitated by the charge nature of electrons, which enables their manipulation via applied electric fields. Quantum computation, therefore, may be effected by applying voltages through biasing gates located on the wafer surface, in particular, "*A* gates", which control the resonance frequency of the nuclear spin qubits, and "*J* gates", which control the electron-mediated coupling between neighboring nuclear spins. In addition, two other biasing magnetic fields are necessary, namely, a global field B_{ac}, to enable flipping of the nuclear spin at resonance, and a local magnetic field, B, to break the two-fold spin degeneracy of germane to electrons occupying the lowest energy-bound state at the donor, which manifests itself at low temperatures.

The detailed physics of the silicon-based nuclear spin quantum computer is captured by the parameters governing the magnitude of the spin interactions, which determines the time required for manipulating qubits and the separation required between adjacent donors. In the presence of a magnetic field $B \| z$, and assuming a donor nucleus with $I = 1/2$ embedded in a silicon host, the interaction in question, namely, the nuclear-spin interaction, is given by the Hamiltonian [202],

$$H_{e-N} = \mu_B B \sigma_z^e - g_N \mu_N B \sigma_z^N + A \sigma^e \cdot \sigma^N, \tag{37}$$

where μ_N is the nuclear magneton, σ are the Pauli spin matrices, g_N is the nuclear g-factor, and $A = \frac{8}{3} \pi \mu_B g_N \mu_N |\Psi(0)|^2$ is the contact hyperfine interaction energy when the probability density of the electron wavefunction, $|\Psi(0)|^2$ is evaluated at the nucleus. Clearly, examination of Eq. (37) indicates that the interaction energy is a directly proportional to the magnetic field and is a strong function of the wave function probability density at the nucleus. A trade-off exists, however, because for electrons in their ground state the frequency separation between nuclear levels is [202],

$$h v_A = 2 g_N \mu_N B + 2A + \frac{2A^2}{\mu_B B}, \tag{38}$$

which, for fields $B < 3.5T$ is dominated by the second term. Thus, in this regime the magnitudes of the nuclear magneton and the wavefunction probability density at the nucleus take on a dominant character.

To perform arbitrary rotations on the nuclear spin, Kane indicates that it is necessary to alter its precession frequency in comparison with that resulting from the applied magnetic field B_{ac} [202]. This is accomplished by exploiting the fact that the proximity of the donor-nuclear spin system to the *A gate* allows the hyperfine interaction to be reduced by shifting the envelope of the electron-donor wavefunction away from the nucleus, i.e., by reducing $|\Psi(0)|^2$. In essence, such a shifting alters the frequency, Eq. (35), and causes the nuclear spin-donor system to behave as a voltage-controlled oscillator producing, for a donor placed 200 Å under the gate, a tuning parameter of the order of 30 MHz/V [202].

In addition to the single-qubit rotation, the two-qubit C-NOT operation must be implemented in order to enable general quantum computations. In the context of the nuclear spin-donor system, accomplishing this requires developing the ability to induce nuclear-spin exchanges between two nucleus-electron spin systems. The interaction between two such systems is captured by the Hamiltonian [202],

$$H = H(B) + A_1 \sigma^{1N} \cdot \sigma^{2e} + A_2 \sigma^{2N} \cdot \sigma^{2e} + J \sigma^{1e} \cdot \sigma^{2e}, \tag{39}$$

where $H(B)$ represents the magnetic field interaction terms between spins, the respective hyperfine interaction energies of the nucleus-electron systems is given by A_1 and A_2, respectively, and $4J$ is the exchange energy, which is a function of the electronic wavefunction overlap and, for donors in a host semiconductor of dielectric constant ε, and Bohr radius a_B, and separated by a distance *r* of about 100-200 Å, is given by [203],

$$4J(r) \cong 1.6 \frac{e^2}{\varepsilon a_B}\left(\frac{r}{a_B}\right)^{\frac{5}{2}} exp\left(\frac{-2r}{a_B}\right). \tag{40}$$

The wavefunction overlap, to which *J* is proportional, is captured by this exchange energy. Thus, varying the voltage applied via the *J*-gate one can modulate coupling between separated qubits.

Once qubits have been manipulated to effect a quantum computation, the result of the computation must be read off. In the silicon-based nuclear spin QC, this is accomplished by measuring the current that results from the conversion of nuclear spins into electron polarization, in response to a bias voltage, see Fig. 4-20 below. In particular, this conversion of the nuclear spin into an electron polarization is prompted by the coupling of the states $|\downarrow\downarrow\rangle$ and $|\uparrow\downarrow - \downarrow\uparrow\rangle$, which is produced by the hyperfine interaction between the nuclei and the electronic states as the exchange energy J is increased adiabatically from $J < \mu_B B/2$ to $J > \mu_B B/2$, see Fig. 4-20(a) [199].

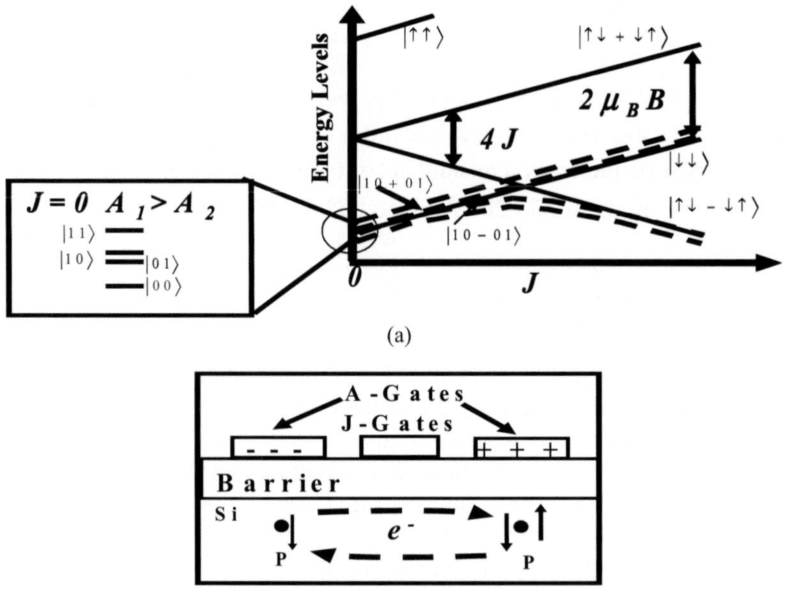

(b)

Figure 4-20. (a) Energy levels for electrons (solid lines) and lowest energy-coupled electron-nuclear (dashed lines) systems as a function of exchange energy, *J*. When $J < \mu_B B/2$, it is possible to perform two-qubit computations by exercising control over the level splitting $|10-01\rangle - |10+01\rangle$ with the J-gate. Above $J = \mu_B B/2$, the states of the coupled system evolve into states with differing electron spin polarization. When $J = 0$ the state of the nucleus with the larger energy splitting, which is controllable by the A-gate, determines the final electron spin state after an adiabatic increase in J. (b) Only electrons in state $|\uparrow\downarrow - \downarrow\uparrow\rangle$ can make transitions into states in which electrons are bound to the same donor (D^- states). These transitions elicit an electron current that is measurable by capacitive means, thus enabling the underlying spin states of the electrons and nuclei to be determined. [202].

This implies a change in wavefunction symmetry, i.e., from that of $|\downarrow\downarrow\rangle$ to that of $|\uparrow\downarrow - \downarrow\uparrow\rangle$.

Two electrons with the latter symmetry, however, are capable of occupying the same donor. In the Si:P the donor takes the form of a D^- state, which is always a singlet state with a second electron binding energy of 1.7meV. Under these circumstances, it will be possible, with the appropriate bias between the A-gates, to induce electrons from one donor to move the adjacent, already occupied one in order to establish the D^- state in it. This charge motion, in turn, is detectable utilizing single-electron capacitance techniques and produces a signal that remains observable until the spin relaxes; for Si:P this time may be of the order of hours [202].

Kane points out that a number of practical considerations must be addressed to make this scheme workable [202]. For instance, before beginning a computation, initialization will require the individual determination of gate biases to account for fluctuations due to the variation with position of both donors and gate sizes. These voltages, in turn, will have to be stored to effect the calibration as needed. Also, gate voltage fluctuations in essence couple the environment to the qubits, thus contributing to spin decoherence. This decoherence is elicited by the induction of difference spin precession frequencies in pertinent qubits, and manifests in that two spins in phase at a given time, will be $180°$ out of phase a time t_ϕ later. It can be shown that [202],

$$t_\phi = \frac{1}{\pi^2 \alpha^2 (V) S_V (\nu_{st})}, \tag{41}$$

where $\alpha = d\Delta/dV$ is the tuning parameter of the A-gates, with Δ the fluctuating differential precession frequency of the spins, S_V is the spectral density of the frequency fluctuations, and ν_{st} is the frequency difference between the $|10-01\rangle$ and $|10+01\rangle$ states. Estimates, assuming the use of low-temperature elctronics to bias the gates, suggest $t_\phi \approx 10^6$ sec, which implies the ability of the nuclear spin QC to perform between $10^5 - 10^{10}$ logical operations during t_ϕ. Finally, measures have to be taken to render a predominance of certain polarization of electrons spins, e.g., $(n_\uparrow / n_\downarrow < 10^{-6})$, so that they can effectively mediate nuclear spin interactions. This, in turn, requires the electrons to occupy the lowest energy levels, which occurs when $2\mu_B B >> kT$. With $B \approx 2T$, this sets the operating temperature at 100mK.

4.3.1.4 Superconducting-Based Qubits

In the search for two-level quantum systems upon which qubits might be based, Josephson junction-based superconducting qubits are currently the most advanced. In contrast to the previously discussed qubits, which are based on microscopic quantum effects of individual particles, such as ions, electrons, or nuclei, superconducting-based qubits are based on *macroscopic quantum coherence* effects [204], [205]. These are effects in which the qubit state is embodied, not in the wavefunction of elemental particles, but on the coherent collective behavior of many particles, e.g., a superfluid. Thus, the qubit states are defined by macroscopically observed quantities, such as the charge or the current of particle condensates.

The key to superconducting qubits is the *nonlinear* nature of the resonant LC circuit embodied in the Josephson junction [206]. The quantum mechanical behavior of a linear LC circuit is captured by the flux Φ through the inductor, which plays the role of position coordinate, and the charge Q on the capacitor, which plays the role of conjugate momentum, thus enabling the commutation relation $[\Phi, Q] = i\hbar$. With the Hamiltonian given by, $H = \Phi^2/2L + Q^2/2C$, the usual eigenenergy states are given by $E = \hbar\omega_0(n + 1/2)$, where $\omega_0 = 1/\sqrt{LC}$ is the resonance frequency. Reflecting the quadratic nature of the potential, the energy states are equally spaced. Thus, it is difficult to define the two lowest states as the qubit states, since transitions between higher-lying states are as equally likely [206].

The LC resonator may be made useful as a qubit if its energy spectrum is caused to exhibit two lowest-lying states separated from the higher-lying states. This is accomplished if a nonlinear inductance is introduced [206]. In particular, the nonlinear Josephson inductance, $L_J = \Phi_0 / 2\pi I_0 \cos\delta$, where $\delta = \phi_L - \phi_R$, $\phi_{L,R}$ is the phase of the wavefunction on either side of the junction, and I_0 is the critical current, introduces a nonlinear potential in which the two lowest-lying states are well separated from the higher-lying states. These variables afford characterization of the Josephson junction in terms of its energy, $E_J(\Phi_{ext}) = \Phi_0 I_0 \cos\delta / 2\pi = E_J \cos\delta$. In this context, the conjugate variables of the quantum mechanical description of the LC resonator become the flux, now given by $\Phi = \varphi_0\theta$, where $\varphi_0 = \Phi_0/2\pi$, and $\theta = \delta \bmod 2\pi$ represents a point in the unit circle (an angle module 2π), and the charge, now given by $Q = 2eN$, which represents the charge that has tunneled through the junction, and N an operator with integer eigenvalues capturing the number of Cooper pairs that have tunneled. The commutation relation now is given by $[\theta, N] = i$ [206]. The Hamiltonian is given by,

$$H = E_{CJ}(N - Q_r / 2e)^2 - E_J \cos\theta, \tag{42}$$

where $E_{CJ} = (2e)^2/2C_J$ embodies the Coulomb energy for adding one Cooper pair worth of charge to the junction capacitance C_J, and Q_r embodies a residual random charge capturing an initial charge existing on the capacitor before it was connected to the inductor [206]. Q_r originates from the inevitable work function difference and/or the presence of excess charged impurities on the capacitor electrodes of the junction.

In the course of developing approaches to minimize the effect of Q_r, while retaining the nonlinearity of the resonator, three fundamental types of Josephson-based superconducting qubits have been developed, namely, the charge qubit, the flux qubit, and the phase qubit, see Fig. 4-21.

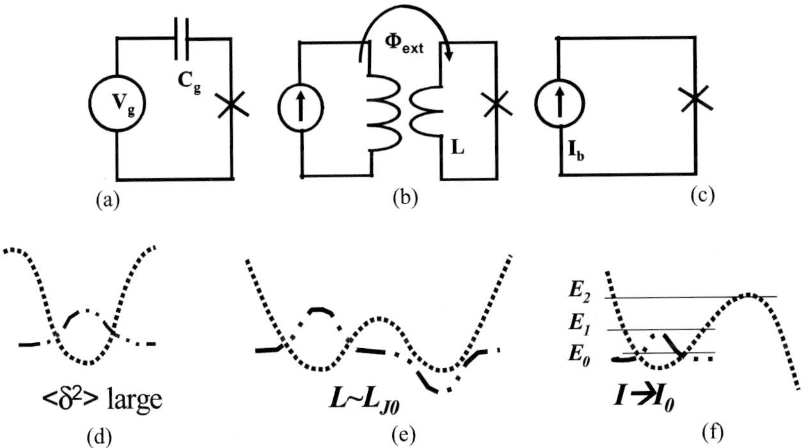

Figure 4-21. Fundamental types of superconducting qubits. (a) Charge qubit. (b) Flux qubit. (c) Phase qubit. (d), (e), (f) Potential (dotted line), showing qualitatively different shapes for these three respective qubit types. In (e) the nonlinearity of the first levels comes about from the cancellation between the superconducting loop inductance and the junction inductance near $\Phi_{ext} = \Phi_0 / 2$. No closed-form expressions exist for the eigenvalues and eigenfunctions of the potential, but its features are captured by two aspect ratios, namely, E_J / E_{CJ} and $\lambda = L_J / L - 1$. Ground-state wavefunction is also indicated (dashed-double-dot line). The "x" represents a Josephson junction. (*After* [206] and [207].)

The nature of the Josephson-based qubit is a function of the relationship between the relative magnitudes of the Josephson energy, E_J, which reflects the strength of the coupling across the junction, and the Coulomb charging energy, E_{CJ}, which reflects the energy needed to increase the charge on the junction by a Cooper pair, $2e$ [208].

4.3.1.4.1 The Charge Qubit

The charge qubit, see Fig. 4-22, also known as the *Cooper pair box*, aims at compensating the residual offset charge Q_r by biasing the Josephson junction with a voltage source V_g in series with a "gate" capacitor C_g. In this case it can be shown that the Hamiltonian, with potential shown in Fig. 4-21(d), is given by,

$$H = E_C(N - N_g)^2 - E_J \cos\theta \ , \tag{43}$$

where $E_C = (2e)^2/(2(C_J + C_g))$ represents the energy required for charging the island of the box and $N_g = Q_r + C_g V_g / 2e$. To function as a charge qubit, $E_{CJ} > E_J$, in which case the circuit favors fixing the numbers of Cooper pairs. In the absence of tunneling, this state of affairs yields an energy versus gate voltage as given by the dashed lines in Fig. 4-22(b), that is, as the gate voltage increases, the energy of the zero state $|0\rangle$ increases and that of the one state $|1\rangle$ decreases.

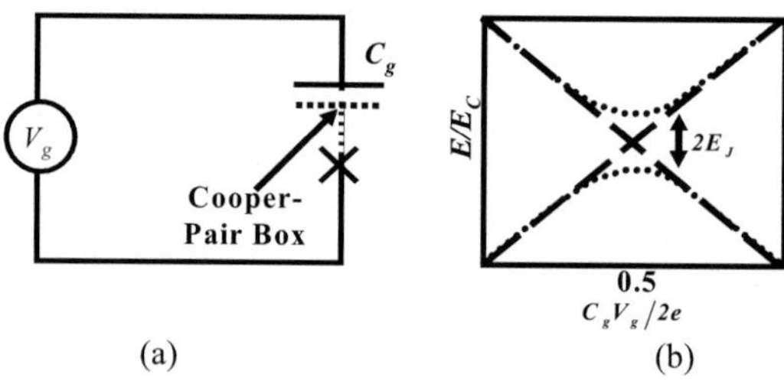

Figure 4-22. Charge qubit. (a) A qubit is created by the superposition of the two classical states embodied by the presence of zero and one extra Cooper pair in the box. (b) Energy levels as a function of controlling gate voltage.

However, in the presence of tunneling, coupling causes the energy levels to split and avoid crossing, thus reflecting the creation of two new quantum states (solid lines), namely, one materialized as the symmetric superposition of the classical zero and one states $(|0\rangle + |1\rangle)$, and the other as their antisymmetric superposition $(|0\rangle - |1\rangle)$, both separated by an energy gap of magnitude $2E_J$ [208].

The dynamic behavior of the charge qubit is controlled by applying time-varying signals to the voltage gate. Initial demonstration of the coherent control of macroscopic quantum states in a single-Cooper-pair box was reported by Nakamura, *et al.* [209]. In these experiments, the superposition of two charge states (i.e. states with different number of Cooper pairs N) was detected by a tunneling current through a probe junction. In particular, a normal electron escaped through the probe junction every time the system adopted the one state. Control of the state of the qubit was effected by varying the length of the voltage pulse, with the probability of the system returning to the zero or one state oscillating in proportion to it. The major source of decoherence was found to be the probe junction itself, which limited the coherence time to 2 ns [206].

Nakamura *et al.'s* [206] approach was improved by the *quantronium* device demonstrated by Vion *et al.'s* [112] see Fig. 4-23. In this device, the Josephson junction of the Cooper pair is split into two small parallel Josephson junctions which are characterized by their energy $E_J \cos(\delta/2)$, where δ is the superconducting phase difference across the series combination of the two junctions. These junctions, in turn, are shunted by a larger Josephson junction, characterized by an energy $E_{J0} \approx 20E_J$ and by a phase γ, thus forming a loop. A current I_ϕ applied to an adjacent coil produces a flux Φ that passes through the loop, with the consequence that it induces a phase ϕ that now links the loop phases as follows, $\delta = \gamma + \phi$, where $\phi = 2e\Phi/\hbar$. This action entangles the state of the box, N, via δ, with the phase γ, see Fig. 4-23(a). The quantum state of the qubit is manipulated by applying a microwave pulse of frequency $\nu \cong \nu_{01} \sim 16.5\text{GHz}$, the transition frequency between charge levels in the box corresponding to the zero and one states. Depending on the pulse duration, any state $|\Psi\rangle = \alpha|0\rangle + \beta|1\rangle$ can be prepared. Reading the state exploits the fact that a current pulse $I_b(t)$, see Fig. 4-23(b), of peak amplitude slightly below the critical current of the large junction, $I_0 = 2eE_{J0}/\hbar$, causes a supercurrent to develop in the loop that is proportional to N. In particular, when there is no extra charge in the box, this supercurrent elicits a clockwise current in the loop formed by the two junctions, whereas when there is an extra charge in the box, the current is counterclockwise. In the former case, the current adds to the bias current in the large junction with the result that, for precisely adjusted amplitude and duration of the $I_b(t)$ pulse, it switches to a finite voltage for a state one and

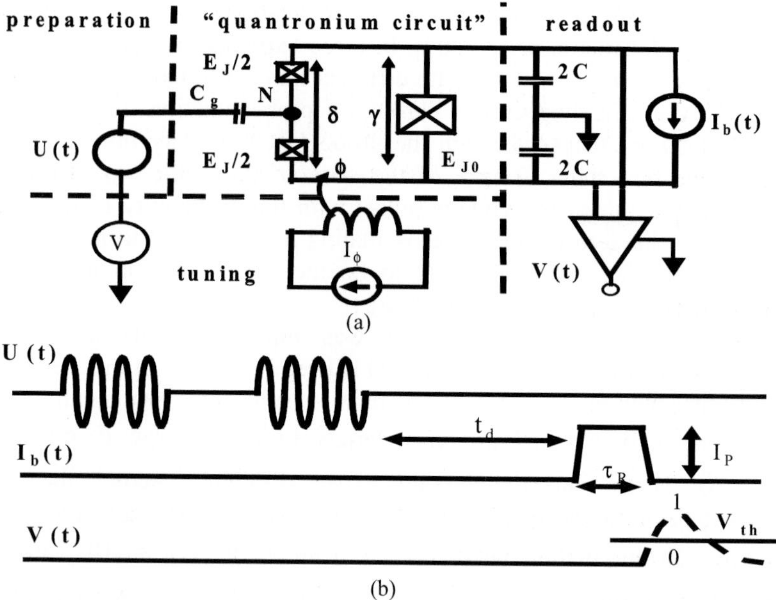

Figure 4-23. Quantronium circuit. (a) The circuit consists of a Cooper pair box island (node N), to which two small Josephson junction branches are connected. These, together with a larger Josephson junction, that is shunted by a capacitance C (to reduce phase fluctuations), form a loop. The state of the circuit is embodied by the number of Cooper pairs, N, and the phases δ and γ. To tune the quantum energy levels, a DC voltage V is applied to the gate capacitance, C_g, and a DC current I_ϕ is forced through the coil to produce a flux ϕ in the circuit loop. (b) To prepare arbitrary quantum states, microwave pulses $U(t)$ are applied to the gate. To read out the state a current pulse $I_b(t)$ is applied to the large junction and the resulting voltage $V(t)$ across it is measured. A typical write/read timing sequence is shown. (*After* [112].)

it does not switch for a state zero. In essence, the quantronium uses a phase circuit to measure current, instead of the charge, thus avoiding the probe-induced decoherence problem of Nakamura et al's. A decoherence time of $0.5\mu s$ was measured [112].

4.3.1.4.2 The Flux Qubit

The flux qubit, see Fig. 4-21(b) above, is considered as the dual of the charge qubit [206]. It consists of a junction that is coupled to a current source via a transformer, instead of a gate capacitor, with the junction itself being connected in series with an inductance L, and the system being biased by an external flux Φ_{ext} through an auxiliary coil. In the flux qubit the approach to compensating the detrimental effect of Q_r relies on shunting the junction with the superconducting wire of the loop and choosing the condition $E_{CJ} < E_J$. This results in making the quantum fluctuations of q much larger than those of ΔQ_r. The Hamiltonian, with potential shown in Fig. 4-21(e), is given by,

$$H = \frac{q^2}{2C_J} + \frac{\phi^2}{2L} - E_J \cos\left[\frac{2e}{\hbar}(\phi - \phi_{ext})\right], \tag{44}$$

where ϕ is the integral of the voltage across the inductor L, which gives the flux through the superconducting loop, and q is its conjugate variable, which represents the charge on the junction capacitance C_J. Both obey the commutation relation $[\phi, q] = i\hbar$. The prototypical flux qubit consists of three Josephson junctions forming a loop and being controlled by an applied magnetic field perpendicular to the loop to control the phase, see Fig. 4-24.

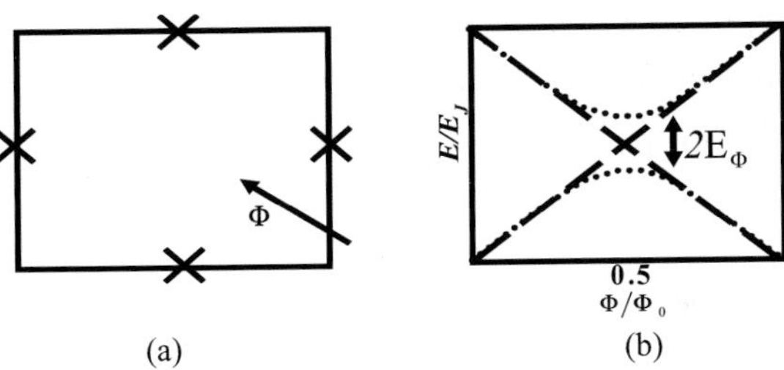

Figure 4-24. Flux qubit. (a) A qubit is created by the superposition of the two classical states embodied by the loop phase of zero and 2π. While one or two junctions would be sufficient, three junctions allow greater control over the behavior of the system. (b) Energy levels as a function of controlling magnetic flux. The energy gap, $E_\Phi = \zeta(\Phi_0^2 / 2L)(N_\Phi - 1/2)$, plays the same role as E_J. ζ is a numerically determined parameter and $N_\Phi = \Phi_{ext} / \Phi_0$. [207], [208].

In this case the two qubit states $|0\rangle$ and $|1\rangle$ are embodied in transitions in phase from loop phases of 0 to 2π, which are associated with currents circulating around the loop in clockwise and anti-clockwise directions. In particular, states of zero and 2π phase difference around the loop, are "coupled" when the flux through the loop equals half the quantum magnetic flux in the superconductor, i.e., when $\Phi = \Phi_0 / 2$. Under this state of affairs, two new states, $(|0\rangle+|1\rangle)$ and $(|0\rangle-|1\rangle)$, that are quantum superpositions, are formed, with the energy between them now given by the tunneling strength. Control of the qubit, such as to change its state, is effected by coupling to the flux ϕ, which is accomplished by sending current pulses on the transformer primary. Measurements of the states, made with a superconducting quantum interference device (SQUID), a device which consists of two Josephson junction in parallel, to detect the magnetic flux, reveals that the currents are carried by a billion Cooper pairs, with tunneling being the mechanism by which the directions of all of these particles is reversed simultaneously [208]. The decoherence times, which are limited by defects in the junction are in the range of 500 ns to $4\mu s$.

4.3.1.4.3 The Phase Qubit

The phase qubit, see Fig. 4-21(c), utilizes only one Josephson junction, and the two quantum states are embodied in the quantum oscillations of the phase difference between junction electrodes [207]. In this case the approach to compensating the detrimental effect of Q_r relies on using large ratios of E_J / E_{CJ}. A large nonlinearity in the Josephson inductance is achieved by biasing the junction at a current $I \sim I_0$. The Hamiltonian, with potential shown in Fig. 4-21(f), is given by,

$$H = E_{CJ} p^2 - I\varphi_0 \delta - I_0 \varphi_0 \cos\delta . \tag{45}$$

The conjugate variables, given by the phase difference operator δ, which is proportional to the flux across C_J, and the charge on the capacitance $2ep$, obey the commutation relation $[\delta, p] = i$ [207]. The potential is approximated by the cubic form,

$$V(\delta) = \varphi_0 (I_0 - I)(\delta - \pi/2) - \frac{I_0 \varphi_0}{6} (\delta - \pi/2)^3 , \tag{46}$$

from where it can be shown that the classical frequency of oscillation at the bottom of the well is given by,

$$\omega_p = \frac{1}{\sqrt{L_{J0} C_J}} \left[1 - (I/I_0)^2\right]^{1/4} , \tag{47}$$

and the first two levels that can be used for the qubit states have the transition frequency $\omega_{01} \cong 0.95\omega_p$ [207].

Read out of the qubit state is accomplished by exploiting tunneling through the barrier separating the potential well from the continuum, and subsequent self-amplification due to the negative slope potential, see Fig. 4-21(f). In particular, since the barrier becomes thinner at higher energies, and those higher energy states have an increasing probability of escape, the one state is measured by sending a probe signal to induce a particle in the one state to tunnel out of the well. Upon tunneling out of the well, the downward acceleration of the potential leads to the appearance of a voltage $2\Delta / e$ across the junction. This voltage is associated with reading a one state for the qubit; zero voltage is associated with reading a zero state.

In terms of operating temperature, it is clear that superconducting qubits must be operated at temperatures such that $kT << \hbar\omega_{01} << \Delta$, where ω_{01} is the transition frequency between the energy levels representing states $|0\rangle$ and $|1\rangle$, and Δ is the energy gap of the superconducting material. This necessitates cooling to temperatures of the order of 20mK.

4.4 Summary

This chapter has dealt with a number of aspects surrounding the actual implementation of NanoMEMS circuits and systems. We began discussing architectural issues, as this is the first step in defining a NanoMEMS system on chip (SoC). Then, emerging candidate building blocks, intended for applications ranging from interfaces to signal processing functions, were described. These included a charge detector, which-path electron interferometer, torsional MEM resonator for parametric amplification, Casimir effect oscillator, magnetomechanically actuated beam, functional arrays, and a quantum entanglement generator. These building blocks represented nanoelectromechanical quantum circuits and systems (NEMX), as they exploited the coexistence of electronic and mechanical structures. The chapter concluded with a presentation of physical implementations of quantum bits (qubits), such as the ion-trap, the nuclear magnetic resonance, the semiconductor solid-state, and superconducting qubits, upon which quantum computing paradigms might be predicated.

Chapter 5

NANOMEMS APPLICATIONS: PHOTONICS

5.1 Introduction

The ability to fabricate nanometer-scale structures has given new impetus to the field of miniaturization of optical devices, whose ultimate goal might be articulated as that of integrating optics and electronics in the context of a monolithic technology. While there are no fundamental limits to the miniaturization of electronic functions down to nano- and sub-nanometer scales, the minimum size of devices manipulating optical signals is limited by diffraction to about half the wavelength ($\lambda/2n$) [210], which in practical terms encompasses dimensions in the several hundreds of nanometer [211]. Two approaches have been devised to overcome these limitations, namely, the design of optical elements based on very high refractive index materials [212], which is accompanied by high losses in the sub-30 nm size regime [213], and the conversion of photons into electromagnetic modes whose size is determined by the size of the waveguide rather than by the wavelength of the optical field [214]. The latter approach is based on surface plasmons (SPs), collective oscillations of free electrons resulting from the interaction of electromagnetic waves with free electrons at a dielectric-metal interface [215]. In particular, Dickson and Lyon [212] point out that, by employing SPs to transport light, the minimum waveguide size becomes only limited by a combination of the Thomas-Fermi screening length, which is ~0.1 nm in Au, and size effects affecting the dielectric constant, which have an onset at dimensions less than 5 nm in Au. While we will focus on SP-based approaches, a third approach to sub-wavelength photonic circuit elements, proposed by Barrelet, Greytak, and Lieber [216], employs semiconducting nanowires and will be touched upon briefly.

In this chapter, we deal with the fundamental principles of *nanophotonics*, the processing of light by nanometer-scale devices. In particular, we address the topics of generation, propagation, and detection of surface plasmons, and emerging devices based on them.

5.2 Surface Plasmons

The concept of plasmons emerges from considering the motion of a concentration $n(\vec{r},t)$ of free electrons, in a positive background n_0, as a result of an applied electric field $\vec{E}$. In particular, assuming the electrons to behave as a fluid of velocity $v(\vec{r},t)$, their motion is prescribed by the consistent solution of Newton's and the continuity equations [132],

$$m\frac{d\vec{v}}{dt}+m(\vec{v}\cdot\nabla)\vec{v}=-e\vec{E}, \tag{1}$$

and

$$\frac{\partial n}{\partial t}+\nabla\cdot(n\vec{v})=0. \tag{2}$$

As a first step towards the solution, after neglecting the second term in (1) due to its quadratic nature in $\vec{v}$, one postulates that the effect of the electric field is to cause the local electron density to deviate from the constant background density by $\delta n = n - n_0$. In this context, the extent of this deviation is related to the electric field by Poisson's equation,

$$\nabla\cdot\vec{E}=-4\pi e(n-n_0)=-4\pi e\delta n, \tag{3}$$

and, because of electron inertia and the restoring force supplied by Coulomb attraction to regain equilibrium, i.e., $\delta n = 0$, oscillations ensue. These collective bulk electron oscillations are denoted as volume *plasmons*, and their frequency of oscillation is obtained by substitution of δn into (2), resulting in,

$$\frac{\partial \delta n}{\partial t}+n_0\nabla\cdot\vec{v}=0, \tag{4}$$

which, upon differentiating with respect to time, becomes,

$$\frac{\partial^2 \delta n}{\partial t^2} + n_0 \nabla \cdot \frac{\partial}{\partial t} \vec{v} = \frac{\partial^2 \delta n}{\partial t^2} + n_0 \nabla \cdot \left(\frac{-e\vec{E}}{m} \right) = 0, \tag{5}$$

and which, in turn, upon substituting (3) into (5) becomes,

$$\frac{\partial^2 \delta n}{\partial t^2} + \frac{4\pi e^2 n_0 \delta n}{m} = 0. \tag{6}$$

Eq. (6), being analogous to that of a harmonic oscillator, prescribes the frequency of plasmon oscillation as,

$$\omega_p = \sqrt{\frac{4\pi n_0 e^2}{m}}. \tag{7}$$

Of particular interest in this chapter, is the concept of *surface plasmons*, (SPs), Fig. 5-1, thoroughly reviewed by Raether [215]. These are elicited by the interaction of external electromagnetic fields with surface electrons, and are characterized by a dispersion relation, a spatial extension, and a propagation length or lifetime.

5.2.1 Surface Plasmon Characteristics

The dispersion relation for SPs at the interface between a dielectric characterized by ε_2, deposited on the plane surface of a semi-infinite metal characterized by $\varepsilon_1 = \varepsilon_1' + i\varepsilon_1''$, is given by [215],

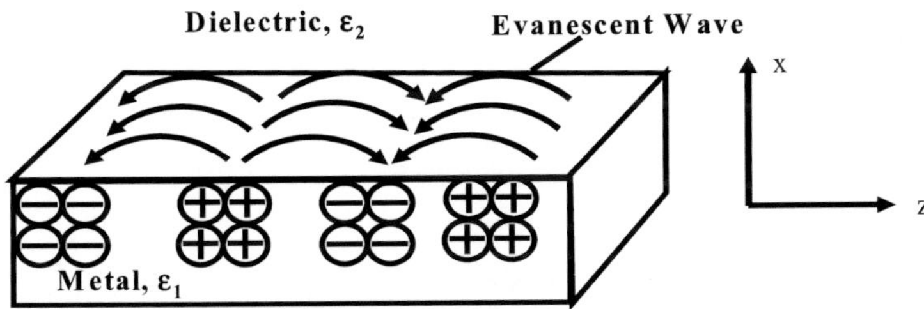

Figure 5-1. Sketch of surface plasmon. The field accompanying a surface plasmon peaks at the dielectric-metal interface and diminishes exponentially away from the interface.

$$k_{zi} = \sqrt{\varepsilon_i \left(\frac{\omega}{c} \right)^2 - k_x^2}, \quad i = 1,2, \tag{8}$$

where the wave vector k_x, is given by,

$$k_x = \frac{\omega}{c}\sqrt{\frac{\varepsilon_1\varepsilon_2}{\varepsilon_1+\varepsilon_2}}. \tag{9}$$

Substituting the complex dielectric constant expression into (9), the wave vector becomes $k_x = k_x' + ik_x''$, with components,

$$k_x' = \frac{\omega}{c}\sqrt{\frac{\varepsilon_1'\varepsilon_2}{\varepsilon_1'+\varepsilon_2}}, \tag{10}$$

and

$$k_x'' = \frac{\omega}{c}\left(\frac{\varepsilon_1'\varepsilon_2}{\varepsilon_1'+\varepsilon_2}\right)^{3/2}\frac{\varepsilon_1''}{2(\varepsilon_1')^2}. \tag{11}$$

Since $\omega/c < k_x$, see (9), and $\varepsilon_1' < 0$, (characteristic of the metal), both k_{z1} and k_{z2} are imaginary. As a result, the SP field becomes evanescent. The corresponding spatial decay of the field, away from the interface, is thus proportional to $\exp(-|k_{zi}||z|)$ [215], and is characterized by the distance at which it has decreased into either medium by 1/e [215]. Thus is given by,

$$z_2 = \frac{\lambda}{2\pi}\sqrt{\frac{\varepsilon_1'+\varepsilon_2}{\varepsilon_2^2}}, \tag{12}$$

into the medium with ε_2, and,

$$z_1 = \frac{\lambda}{2\pi}\sqrt{\frac{\varepsilon_1'+\varepsilon_2}{\varepsilon_1'^2}}, \tag{13}$$

into the medium with ε_1.

The propagation length L_i for SPs propagating along a smooth surface is defined as the distance, away from the interface, at which their intensity, which is proportional to $\exp(-2k_x''x)$, has decreased by 1/e, namely,

$$L_i = \frac{1}{2k_x''}. \tag{14}$$

Raether [215] has pointed out that at visible wavelengths in silver, L_i may be as high as $22\mu m$ at $\lambda = 5145\mathring{A}$, and $500\mu m$ at $\lambda = 10{,}600\mathring{A}$. In addition to characterizing the SP decay by a distance, it may also be characterized by its lifetime. This is related to the SP group velocity by, $T_i = L_i / v_g$ and, is a complex frequency $\omega = \omega' - i\omega''$ and real k_x' are assumed, may be expressed as $T_i = 2\pi/\omega''$, where from (9), one obtains,

$$\omega'' = k_x' c \frac{\varepsilon_1''}{2(\varepsilon_1')^2} \frac{\varepsilon_1' \varepsilon_2}{\varepsilon_1' + \varepsilon_2}. \tag{15}$$

Since SPs are associated with both a field and electron motion, their lifetime is influenced by mechanisms giving rise to attenuation. These include, radiation damping (conversion of the SP into light due to scattering), electron scattering processes giving rise to ohmic losses, and chemical interface damping due to high interface state densities [217]. Two steps are essential, therefore, in the miniaturization of optics by exploiting SPs, namely, the processes of exciting the SPs by light, and of transporting SPs with minimum loss. These subjects are taken up by nanophotonics.

5.3 Nanophotonics

Nanophotonics deals with the realization of nanometer-scale optical components and signal processing functions. While the goal is to produce miniaturized optical components, it is conceivable that components in the SP domain, while performing equivalent optical functions, might take different forms not derivable from a direct downscaling of their optical counterparts. Nevertheless, functions such as light-to-SP conversion, SP wave guiding, and SP-to-light conversion are expected to be fundamental to these pursuits.

5.3.1 Light-Surface Plasmon Transformation

Schemes for converting light into SPs, and vice versa, derive from circumventing the incompatibility of their dispersion relations, which do not intersect, see Fig. 5-2 below, and the necessity to conserve momentum. Accordingly, there are two fundamental elements to supply the additional momentum, namely, the grating coupler, and the ATR prism.

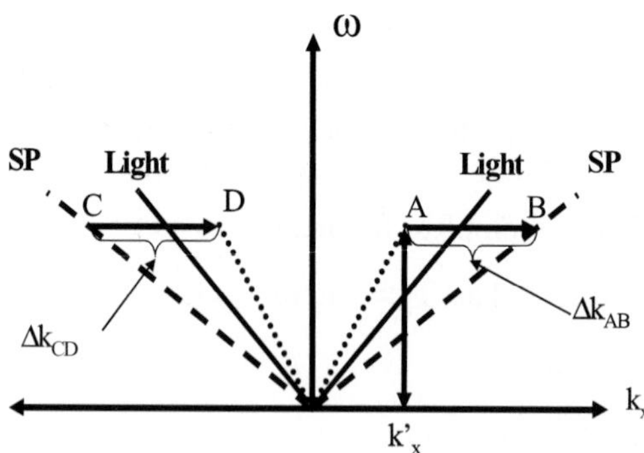

Figure 5-2. Sketch of dispersion relations for light, $k_x = \omega/c$, and SPs, $k_x = \omega\sqrt{\varepsilon_1\varepsilon_2}/c\sqrt{\varepsilon_1+\varepsilon_2}$. An incoming light wave with wave vector k'_x, necessitates and added momentum Δk_{AB} to convert to an SP. Conversely, an SP Necessitates losing a momentum Δk_{CD} to transform to a light wave. (*After* [215].)

In the grating coupler technique, the wave vector of light impinging upon the grating-metal interface at an angle θ is resolved into one component perpendicular to the grating-metal interface, and one component along the interface, see Fig. 5-3. In particular, for a grating of period **a**, the wave vectors along the interface are given by $\omega/c \sin\theta \pm ng$, where n is an integer and $g = 2\pi/a$ is the reciprocal lattice vector of the grating. Coupling between the light and the SPs is achieved when the condition,

$$k_x = \frac{\omega}{c}\sin\theta_0 \pm \Delta k_x = \frac{\omega}{c}\sqrt{\frac{\varepsilon}{\varepsilon+1}} = k_{SP}, \quad (16)$$

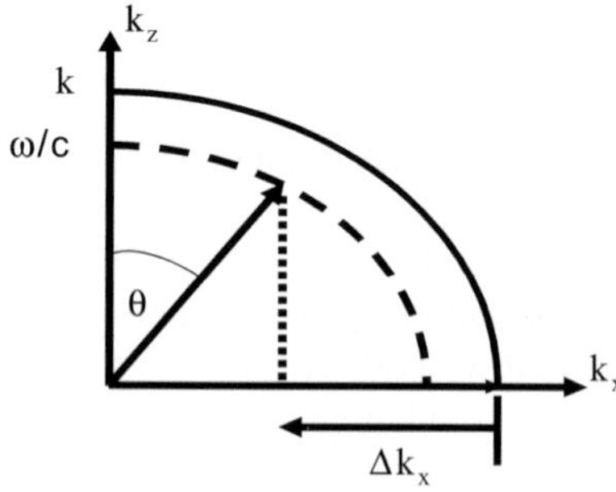

Fig. 5-3 Concept of grating coupler to transform light into SPs. (*After* [215].)

that is, when an incidence angle θ_0 exists at which the sum or difference of the component of the light wave vector and a multiple of the grating reciprocal lattice vector equal an SP wave vector. Reduction of an SP vector by Δk_x transforms it into light, whereas addition of Δk_x to the light's wave vector transforms it into an SP.

In the ATR method, , see Fig. 5-4, the wave vector of light impinging upon a hemispherical prism of dielectric constant ε_0 and the metal interface at an angle θ_0 resolves its wave vector into components that are perpendicular and parallel to the prism-metal interface. In this case, coupling between light and SPs occurs when the component of the light's wave vector along the interface, $k_x = \sqrt{\varepsilon_0}\ \omega \sin\theta_0 / c$, equals the SP wave vector, $k_{SP} = \omega\sqrt{\varepsilon_1\varepsilon_2} / c\sqrt{\varepsilon_1 + \varepsilon_2}$. If the metal thickness is finite, e.g., of extent *d*, there exists the possibility that for a certain value of d, the evanescent field at the $\varepsilon_0 / \varepsilon_1$ interface may couple to the lower $\varepsilon_1 / \varepsilon_2$ interface, where it could also excite SPs [215], see Fig. 5-14.

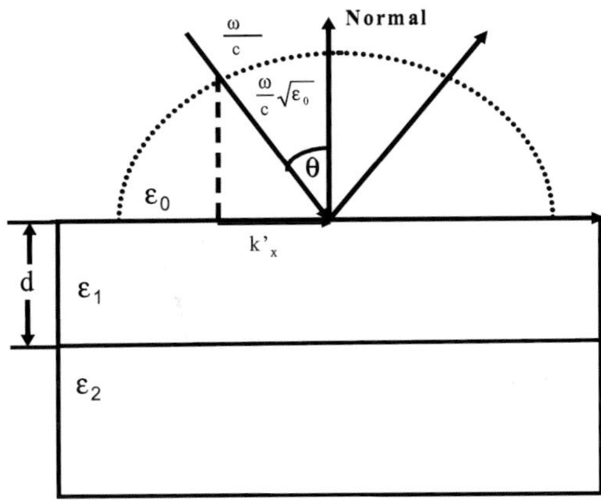

Figure 5-4. Concept of ATR coupler. A metal layer of thickness *d* and dielectric constant ε_1 is sandwiched between a prism of dielectric constant ε_0 and a dielectric ε_2. (*After* [215].)

5.3.2 One-Dimensional Surface Plasmon Propagation

Once light has been converted into SPs, the next question is how to provide efficient energy guidance. To elucidate the issues involved, a number of studies on surface plasmon propagation, utilizing various forms of waveguide, have been undertaken.

5.3.2.1 SP Propagation in Narrow Metal Stripes

Lamprecht *et al.* [214] conducted studies of SP propagation in microscale Au and Ag metal stripes of widths in the micrometer range, and determined the effect of film width on SP propagation length, see Fig. 5-5.

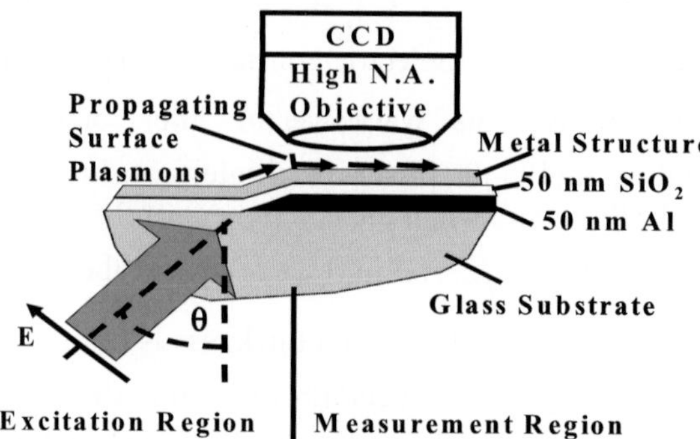

Figure 5-5. Sketch of setup for spatially confined SP excitation and measurement. (*After* [214].)

In particular, they fabricated 70 nm-thick gold and silver stripes with widths in the $54-1\mu m$ range. Their experimental scheme, see Fig. 5-5, involved localized light-SP coupling by a prism arrangement utilizing an opaque aluminum screen to achieve well demarcated excitation and propagation regions. The propagation lengths were observed by detecting SP stray light with a CDD camera upon excitation with three different wavelengths, namely, 514, 633, and 785 nm. The experiment concluded that the SP propagation length decreased with decreasing lateral stripe width, the rate of decrease being very dramatic below $20\mu m$, and increased with wavelength. At a wavelength of 633 nm, the propagation length in a silver stripe was about $58\mu m$ and a few microns, for stripe widths of $54\mu m$ and $1\mu m$, respectively.

5.3.2.2 SP Propagation in Nanowires

Dickson and Lyon [212], conducted studies of SP propagation in high-aspect-ratio metal nanostructures and 20 nm diameter, $1-15\mu m$-long Au and Ag rods, observing propagation over distances greater than $10\mu m$ for light wavelengths of 532 nm and 820 nm. In particular, they reported that once the SP propagation is initiated, the SPs are guided down the length of the wire and reemerge from the end as photons via plasmon scattering. In addition, for specific incident excitation wavelength and waveguide composition, they were able to demonstrate unidirectional SP propagation.

5.3.2.3 SP Resonances in Single Metallic Nanoparticles

Further efforts were made to study the confinement of SPs to metallic nanoparticles. Among these, Klar *et al.* [217] reported the measurement of SP resonances in single metallic nanoparticles, and of the homogeneous line shape of their resonance, via photon scanning tunneling microscopy (PSTM) (PSTM detects a signal at the exit of an optical fiber tip that is proportional to the near field.) These SP resonances are known to be determined by the dielectric properties of the medium in which the particles are embedded, and by the size and shape of the particles, and are accompanied by a large resonant enhancement of the local field both inside and near the particle, see Fig. 5-6 [218].

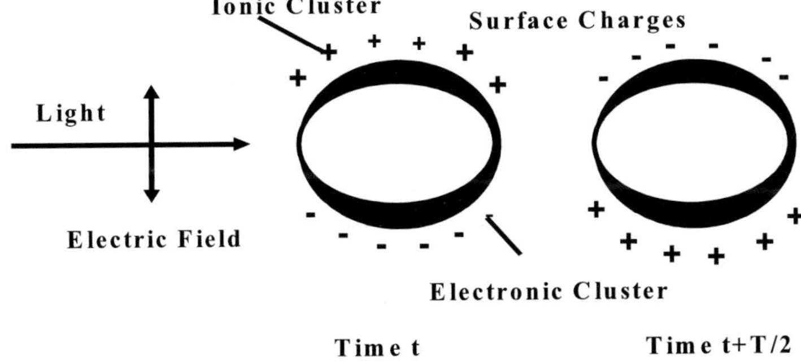

Figure 5-6. Sketch illustrating the excitation of the dipole surface plasmon oscillation. The electric field of an incoming light wave induces a polarization of the free electrons with respect to the much heavier ionic core of a spherical metallic nanoparticle. The net charge difference is only felt at the nanoparticle surface which, in turn, acts as a restoring force. In this way a dipolar oscillation of the electrons is created with period *T*. (*After* [218].)

The setup utilized by Klar *et al.* [217], see Fig. 5-7, consisted of a tunable continuous wave (CW) laser illuminating the sample via a tapered Al-coated fiber tip. The nanoparticles were gold spheres with a typical diameter of 40 nm, and occupying a volume fill fraction of 3 %, embedded in a 200 nm-thick dielectric sol-gel TiO_2 matrix with a refractive index 2.19. The experiment proceeded to position the fiber tip 7 nm from the sample and to shine laser light of various photon energies, in particular, 2,11 eV, 2 eV, 1.94 eV, and 1.91 eV. Detection was effected by a silicon photodetector and plots of the transmitted light intensity, scanned across a surface area of 750 x 750 nm^2 were made. Three key results were obtained in the experiment, namely, an enhanced transmission by a maximum factor of 12, with respect to the background intensity, for a nanoparticle located near the center of the scan area, a typical resonance width of ~160meV, corresponding to a dephasing time of 7fs, and a double-peak resonance structure. The field enhancement was explained as caused by the excitation of the SP resonance by the evanescent field of the fiber aperture and subsequent radiation, by the particle, of propagating modes into the far field, much like an antenna. The

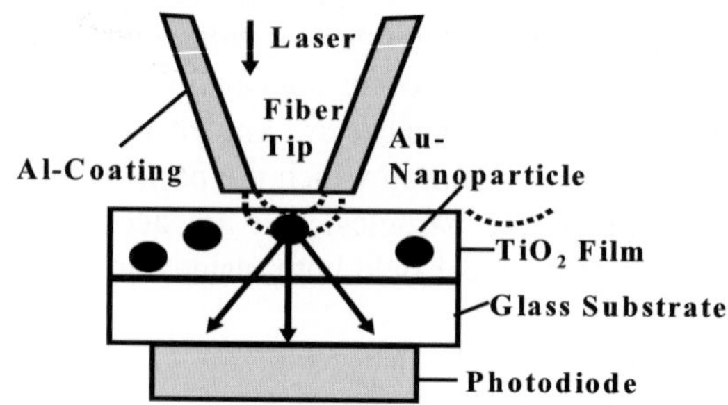

Figure 5-7. Sketch of setup for measuring surface plasmon resonances in single metallic nanoparticles. The fiber tip has an aperture diameter of about 80 nm and positioned 7 nm away from the 200 nm thickTiO_2 film, which is supported by a 1 mm-thick glass substrate. (*After* [217].)

double-peak feature was explained as denoting the electromagnetic coupling of two close-lying particles.

5.3.2.4 SP Coupling of Metallic Nanoparticles

The properties of SP coupling between close-lying metallic nanoparticles were studied by Krenn *et al.* [219] and Kottmann and Martin [220]. Krenn *et al.* [219] utilized PSTM to elucidate the evolution of the optical near-field pattern when a large number of identical particles are arranged in a linear chain. Comparison with theoretical calculations lead them to confirm the unexpected squeezing of the optical near field due to SP coupling above a chain of half oblate Au spheroids nanoparticles with sizes averaging 100 x 100 nm^2 in section, by 40 nm height.

Kottmann and Martin [220] conducted a theoretical investigation of the plasmon resonances of interacting silver cylindrical nanoparticles with 50 nm diameter at various separations, e.g., see Fig. 5-8. This figure shows that at a separation of 5 nm and incidence along the major axis (i.e., along the horizontal arrow) a single cylinder exhibits a resonance (dotted line) at $\lambda = 344\text{nm}$. This resonance has the same magnitude, although shifted down to $\lambda = 340\text{nm}$, for two cylinders (dashed line). In addition, an extra resonance at about 372 nm is observed (dashed line) for this latter case, showing the coupling of the two cylinders. In this case, an enhancement in gap field amplitude, with respect to the incident field amplitude, by a factor of 8 is observed. When the wave is incident normal to the major axis (as indicated by the dashed arrow), a broad resonance is observed at $\lambda = 380\text{nm}$, with a gap field enhancement of 40 with respect to the incident illumination.

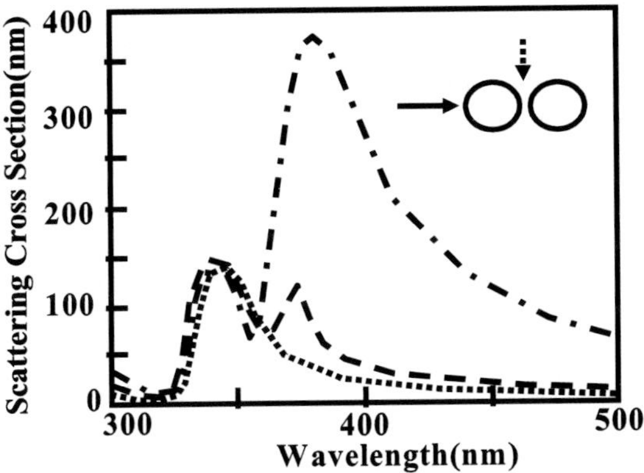

Figure 5-8. Scattering cross section (SCS) calculation of 50 nm diameter cylinders with 5 nm separation. Illumination is in two different directions, as indicated by the arrows in the inset. The incident field polarization is in-plane, perpendicular to the arrows. The dotted curve corresponds to a single cylinder. [220].

5.3.2.5 Plasmonic Waveguides

The concept of exploiting the coupling of resonant SP fields between adjacent metal nanoparticles to realize plasmon waveguides was studied by Maier *et al.* [211] via finite-difference time-domain (FDTD) simulations and experimentally. The FDTD simulations involved exciting a linear array of 50 nm Au spheres with a center-to-center spacing $d = 75\text{nm}$, and driven by a source dipole placed before the first particle. The driving pulse was centered at 2.4 eV, the resonance energy of an individual particle and corresponding to $k = \pi/2d$, the highest group velocity waveguide mode. The pulse had a width of 30 fs, equivalent to 95% of the bandwidth of the dispersion relation for each polarization, and 24% of the total simulation time. For a linear chain of nine nanoparticles, the FDTD simulations predicted group velocities of $1.7\times10^7\,\text{m/s}$ and $5.7\times10^6\,\text{m/s}$ for field excitations of transverse and longitudinal polarization, respectively. Similarly, energy decay lengths, estimated by monitoring the maximum field amplitudes at the center of each particle and at the longitudinally polarized source, of 6dB/280nm and

6dB/86nm were determined. The FDTD study concluded that by optimizing particle geometry it should be possible to achieve energy trabsport at a velocity of 0.1c (c is the speed of light).

The direct experimental evidence of energy transport a waveguide consisting of linear arrays of 90 nm x 30 nm x 30 nm rod-shaped Ag nanoparticles with an inter-particle spacing of 50 nm and having the long axis of the rods oriented perpendicular to the propagation direction to increase the near-field coupling was fabricated. To probe energy transport, the fluorescence of Molecular Probes Fluorspheres F-8801, polystyrene nanospheres with a diameter of $110 \pm 8\text{nm}$, placed randomly along the waveguide, see Fig. 5-9, was detected.

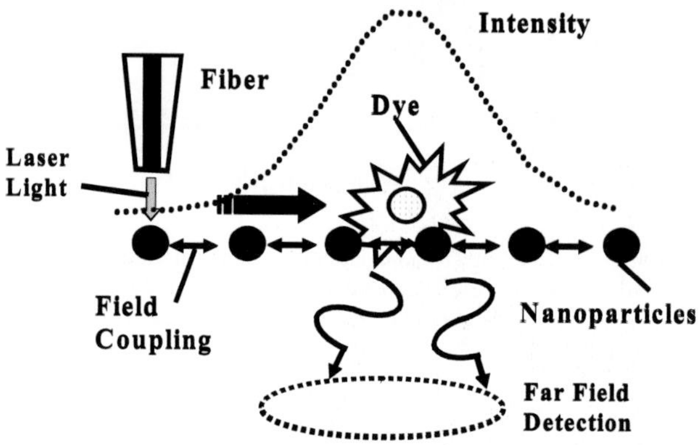

Figure 5-9. Sketch of SP propagation detection along waveguide by fluorescent molecules. (*After* [211].)

The procedure entailed excitation of the first particle in the waveguide by coupling laser light at a wavelength of 570 nm, the single particle resonance wavelength, via the tip of an optical fiber, and monitoring its propagation down the guide by measuring the position-dependent intensity of the light emitted by the fluorescent molecules. The presence of plasmon transport was signaled by a broader full width at half maximum of the fluorescent nano spheres when a scan is done along the waveguide than perpendicular to it. The results of the experiment were a decay length of $6\text{dB} / 195 \pm 28\text{nm}$, corresponding to an energy propagation distance of $0.5\mu\text{m}$.

5.3.3 Nanophotonic SP-Based Devices

While still in its infancy, a number of SP-based devices have been proposed [221], [222]. For instance, Bozhevolnyi *et al.* [221] advanced SP-based waveguiding structures inspired by photonic bandgap crystal (PBC)-based designs. In particular, the propagation of SPs in the range of 780-820 nm launched into nanostructured gold film surfaces with areas of 200-nm-wide scatterers arranged in a 400-nm period triangular lattice containing line defects was demonstrated, see Fig. 5-10.

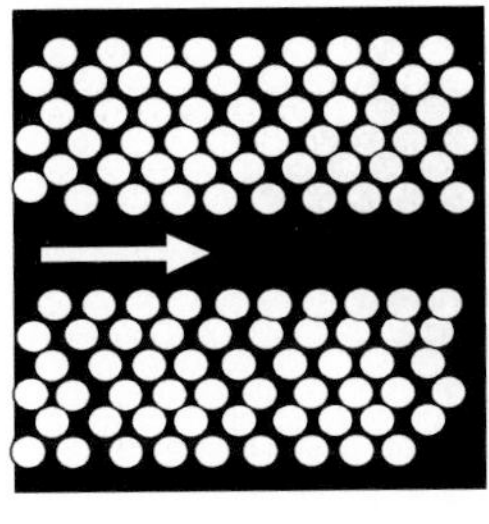

(a)

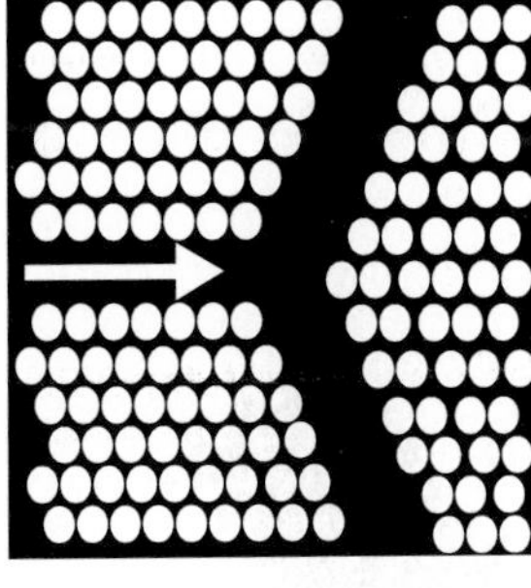

(b)

Figure 5-10. Sketch of SP-PBC devices. (a) Line defect waveguide. (b) Line defect junction. The white circles represent 45-nm-thick gold posts.

The periodicity of the metallic scatterers was arranged to inhibit SP propagation inside these areas, thus creating a plasmonic band gap at a certain range of wavelengths, in particular, at 815 nm. Guidance of SPs occurred at 782 nm along the line defects. This was the first observation of SP band-gaps and SP guiding along line defects in SP-PBC structures. Figure 5-10 shows sketches of the SP-PBCs.

Krenn *et al.* [222], on the other hand, demonstrated two-dimensional optics based on SPs, in particular, local SP sources, Bragg mirrors, and beam interferometer. The goal of the SP source was to launch laterally an SP beam, and was based on the grating approach. In particular, it consisted of periodic nanoscale protrusions on a metallic film with geometries providing the matching between the light and SP wave vectors. SPs were launched by focusing a 750 nm, 5mW laser beam on a silver nanoparticle of 200 nm diameter and 60 nm height. Bragg mirrors, see Fig. 5-11(a), consisted of five

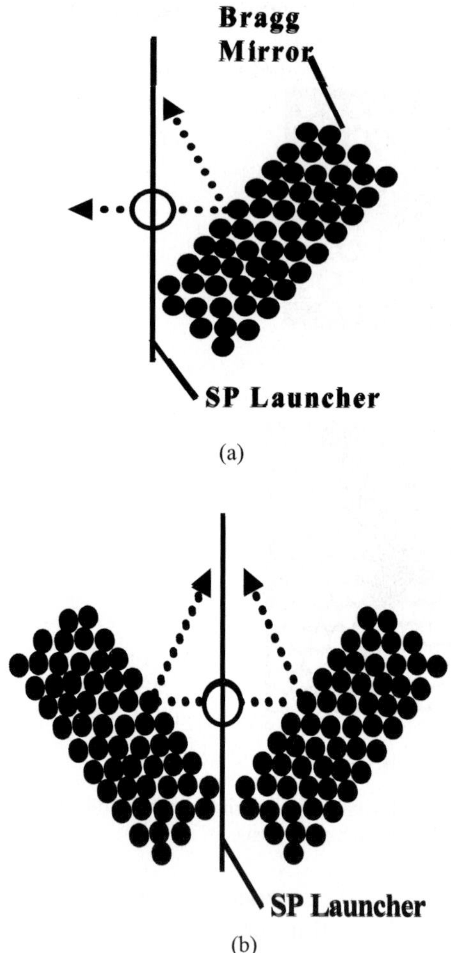

Figure 5-11. Sketches of SP-based devices. (a) Bragg mirror. (b) Beam interferometer. The circle represents the focus of the impinging laser. The dashed arrows represent propagating SPs.

lines of gold 140 nm diameter, 70 nm height gold inclined at a 30° angle with respect to the nanowire used for launching the SPs. Within each line, the center-to-center particle distance was 220 nm and, to fulfill the Bragg condition at an SPP wavelength of 610 nm, the inter-line distance was 350 nm. A reflection coefficient of ~90% was estimated. Since the transmitted intensity was found to be negligible, this was taken to mean that 10% of the SP intensity was converted to light. A beam interferometer was configured by combining two Bragg mirrors symmetrically with respect to a nanowire used for launching SPs, see Fig. 5-11.

5.3.4 Semiconducting Nanowire-Based Nanophotonics

In addition to the SP-based nanophotonics approach, an approach based on using active nanowire waveguides has been advanced by Lieber's group [216]. This approach is motivated by an attempt to circumvent the loss limitations exhibited by passive waveguides, such as SP-based devices, which may hinder their applicability for manipulating light over the extent of integrated photonic systems.

Early examples of semiconducting nanowires include nanoscale lasers [223], in which a sub-wavelength diameter nanocavity is created by exploiting the high refractive index contrast between a nanowire and its surroundings. The active waveguide concept pursued by Lieber's group [216] involves utilizing cavities such as these as waveguides. The feasibility of the concept was investigated by quantitatively characterizing the losses through straight and sharply bent CdS nanowires, of sub-wavelength (200 nm) diameter, by scanning optical microscopy. In particular, the experiments recorded spatial maps of the intensity of light emitted from one end of the nanowire, as a function of the position of a diffraction-limited laser spot with energy greater than the CdS band gap. In this context, the laser energy absorbed by the CdS nanowire was re-emitted via photoluminescence and subsequently guided by it. The experiment indicated that active CdS nanowires are capable of efficient guiding over straight and sharp and acute angle bends, with typical losses of about 1-2dB in an abrupt bend. In addition, by studying the characteristics of junctions between two nanowires it was found that light may be coupled efficiently through sub-wavelength bends defined by them. Finally, by applying a variable electric field across a nanowire, it was demonstrated that it is possible to modulate the intensity of the light exiting the nanowire ~25% at a field of $\sim 2.4\times10^5\,\mathrm{V/cm}$.

5.4 Detection of Surface Plasmons

The detection of SPs relies on their conversion to light, and the subsequent detection of this light. In this context, one can mention two detection schemes. In one scheme, detection is effected by monitoring the light emitted by fluorescent molecules covering the entire device; such was the approach employed in Section 5.3.2.5 to show direct evidence of SP propagation in a plasmon waveguide [211]. This approach is more of a diagnostic tool and does not seem amenable to utilization in actual signal processing systems where one is interested in detecting the output at the exit of, e.g., a nanowire. In a second approach, a near-field scanning optical microscope (NSOM), which allows sub-wavelength resolution [224], is utilized. In this section we provide the fundamental principles of operation of the NSOM.

5.4.1 NSOM/SNOM

Near-field scanning optical microscopy (NSOM), also called scanning near-field optical microscopy (SNOM), is a super-resolution optical microscopy technique that enables the ability to view samples at spatial resolutions beyond those attainable with conventional optical techniques [224], [225]. Conventional optical techniques are limited by the diffraction of light. This is characterized by the size of the spot to which a light beam can be focused. The spot is part of a family of concentric rings, known as the Airy disk pattern, and its size is defined as the distance *d* from the point of highest intensity, located at the middle of the center spot, to the first node in intensity (demarcating the beginning of the first ring), and it is given by,

$$d = 0.61 \frac{\lambda_0}{n \sin \theta}, \tag{17}$$

where λ_0 is the free-space wavelength, n is the index of refraction on the medium in which the light propagates, and θ is the angle describing the light convergence for the focusing element [225]. With the value of the denominator, denoted as numerical aperture (NA), for the objective, being typically as high as 1.3-1.4, (17) is usually simplified to $d = \lambda_0/2$. This is taken as the distance two objects may be approached to one another other while still being distinguishable. To circumvent this limit, Synge [226], [227] proposed the scheme shown in Fig. 5-12.

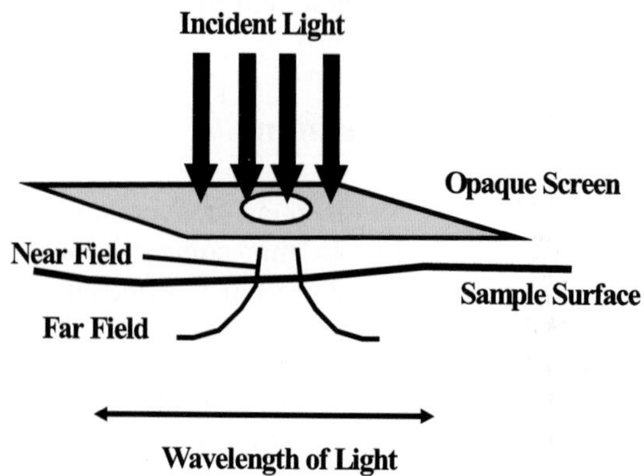

Figure 5-12. Sketch of Synge's concept for overcoming diffraction limit. (*After* [225].)

Here an opaque screen containing an aperture of dimension much smaller than the optical wavelength is interposed in the light path, in front the sample surface, thus circumscribing the passing light to diffract from this small aperture. Fig. 5-13 shows a sketch of a typical SNOM imaging system.

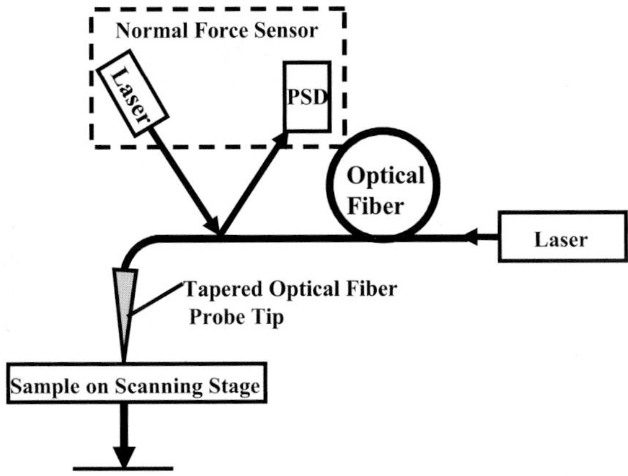

Figure 5-13. Sketch of typical SNOM system. The probe-sample distance is controlled via normal force feedback. (*After* [228].)

By placing the sample surface in the immediate vicinity of the aperture, the light emerging from it would be made to interact with the sample before diffracting out, thus allowing a higher resolution image to be formed. In practice, the sample is illuminated via a 50-100 nm-diameter hole in a tapered optical fiber probe tip [228]. The system may be operated in at least four modes, Fig. 5-14, according to whether the probe tip is used for illumination, for light collection, or for both [228]. In the transmission mode, Fig. 5-14(a), the probe tip illuminates the sample and the transmitted light is collected and processed. In the reflection mode, the probe tip illuminates the sample, and the reflected sample is collected and processed. In the collection mode, Fig. 5-14(c), an external light source illuminates the sample, and the probe tip collects the light reflected from the surface. In the illumination and collection mode, Fig. 5-14(d), the probe tip is employed to both illuminate the sample and collect the reflected light.

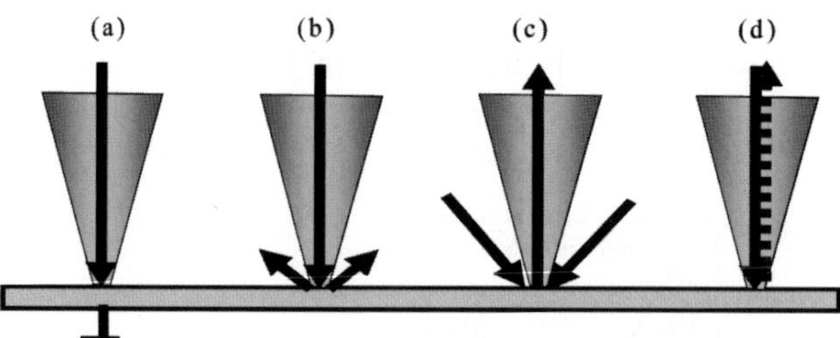

Figure 5-14. Modes of operation of SMON system. (a) Transmission mode imaging. (b) Reflection mode imaging. (c) Collection mode imaging. (d) Illumination/collection mode imaging. (*After* [228].)

The theory of diffraction by small holes was originally treated by Bethe [229] and corrected by Bouwkamp [230], [231]. The proper expressions for the field components in the near-field region in the immediate vicinity of the aperture are given by [231],

$$E_x = ikz - \frac{2ikau}{\pi}\left\{1 + v \cdot \arctan v + \frac{1}{3}\frac{1}{u^2+v^2} + \frac{x^2-y^2}{3a^2\left(u^2+v^2\right)\left(1+v^2\right)^2}\right\}, \quad (18)$$

$$E_y = -\frac{4ikxyu}{3\pi a\left(u^2+v^2\right)\left(1+v^2\right)^2}, \quad (19)$$

$$E_z = \frac{4ikxv}{3\pi a\left(u^2+v^2\right)\left(1+v^2\right)^2}, \quad (20)$$

where **a** is the aperture radius, k is the wave number, and x, y, and z are related to the oblate-spheroidal coordinates u, v, and φ via the equations,

$$x = a\left[\left(1-u^2\right)\left(1+v^2\right)\right]^{1/2}\cos\varphi, \quad (21)$$

$$y = a\left[\left(1-u^2\right)\left(1+v^2\right)\right]^{1/2}\sin\varphi, \quad (22)$$

$$z = auv. \quad (23)$$

5.5 Summary

This chapter has dealt with the application of NanoMEMS techniques to photonics. After pointing out the limitations of conventional optics to render miniaturized devices at sub-wavelength sizes, we went on to consider the paradigm of surface plasmons to enable such miniaturization. In particular, the issues of converting light to surface plasmons, as well as a number of approaches to SP propagation, i.e., in narrow metal stripes and in nanowires, were discussed. Then, the behavior of SPs on nanoparticles was presented, followed by the phenomena of coupling between SPs in adjacent nanoparticles, and its subsequent application to create plasmonic waveguides. An alternate proposal to realize nanophotonics, based on active semiconducting nanowires was then presented. The chapter concluded with a discussion of the near-field scanning optical microscopy technique to detect surface plasmons.

Appendix A

QUANTUM MECHANICS PRIMER

A.1 Introduction

In this appendix we present some of the salient point of quantum mechanics (QM) of relevance to the material in this book. These include the basic laws governing quantum systems, the harmonic oscillator and quantization, creation and annihilation operators, the second quantization formalism,and field operators.

A.2 Some Basic Laws Governing Quantum Systems

Phenomena occurring at microscopic scales is governed by quantum mechanics (QM) [60]. According to QM, all the information regarding a microscopic particle (e.g., momentum and position) is contained in its wavefunction, ψ. This wavefunction obeys an *operator* equation, namely, Schrödinger's equation, and is determined by the total energy of the particle. The possible energy states of the particle are given by the solutions to the *stationary* Schrödinger's equation,

$$\hat{H}\psi = E\psi, \tag{A.1}$$

where $\hat{H}$ is the Hamiltonian operator, which embodies the total energy of the particle and is composed of the sum of its kinetic and potential energies, and E is its eigenvalue. Since the result of measuring energy are real values, $\hat{H} = \hat{H}^{\dagger}$, i.e., $\hat{H}$ is hermitian. In general, there can be a multitude of eigenvalues, both discrete and/or continuous, each one being accompanied by a respective eigenfunction. Thus, the set of wavefunctions associated with an operator are said to span a space, called state space. When a particle is not in a stationary state, it is in a dynamic state. A particle is in a dynamic state when it is between stationary states, and the dynamic state is embodied by a *superposition* of stationary states. During these circumstances, the particle's state is found as a solution to the time-dependent Schrödinger's equation,

$$\hat{H}\psi = i\hbar\frac{\partial\psi}{\partial t}, \tag{A.2}$$

where $\hbar$ is Planck's constant and t is time. If there are n stationary states, then the solution to (A.2) is expressed as,

$$\psi = c_1(t)\psi_1 + c_2(t)\psi_2 + \ldots + c_n(t)\psi_n, \tag{A.3}$$

where the wavefunctions ψ_i correspond to respective stationary states with energies E_i, and for a normalized state $|c_i|^2$ represents the probability that, upon measuring the state of the particle, it will be found in state *i.* A state is normalized when its inner product $(\psi,\psi)=1$. Thus, for a normalized state $\sum_i |c_i|^2 = 1$. But this is the norm of ψ, therefore, the norm of the state vector remains constant, i.e., does not depend on time. Two proportional state vectors, say, ψ' and ψ, where $\psi' = ce^{i\theta}\psi$, represent the same physical state, but in general, the superposition of states possessing expansion coefficients with relative phases, such as $\psi'' = c_1e^{i\theta_1}\psi_1 + c_2e^{i\theta_2}\psi_2$ does not.

The state of a particle deprived of interaction with its environment, will evolve according to the solution to (A.2), which, expressed in Dirac's ket notation, is given by,

$$|\psi(t)\rangle = U(t,t_0)|\psi(t_0)\rangle, \tag{A.4}$$

where, when $\hat{H}$ is time-independent, U is the evolution operator

$$U(t,t_0) = e^{-\frac{i}{\hbar}\int_{t_0}^{t} H dt'} = e^{-iH(t-t_0)/\hbar}. \tag{A.5}$$

A system whose Hamiltonian is time-independent exhibits energy conservation over time, i.e., the total energy is a constant of the motion. Clearly, $U^{\dagger}U = UU^{\dagger} = 1$. This means that U conserves the norm of the states it acts upon, i.e., it is a unitary operator. Also, since $U(t,t_0)=U^{-1}(t_0,t)$, this means the system is *reversible*. When the system is disturbed by (or coupled to) the environment, as a result of which its energy is modified, then its evolution is modified, the norm is no longer conserved, the system becomes *irreversible*, and the state is said to *decohere*.

A.3 Harmonic Oscillator and Quantization

In the simplest case of a particle of mass *m* and constant total energy (Hamiltonian), *H*, performing an oscillatory motion in a potential $V(q)=kq^2/2$, with kinetic energy $T=\frac{1}{2}m\left(\frac{dq}{dt}\right)^2=p^2/2m$, Schrödinger's equation is given by,

$$\hat{H}\psi=\left[\hat{T}(q)+\hat{V}(q)\right]\psi=\left[\frac{\hat{p}^2}{2m}+\hat{V}(q)\right]\psi=\left[-\frac{\hbar^2}{2m}\frac{d^2}{d\hat{q}^2}+\frac{1}{2}m\omega^2\hat{q}^2\right]=\varepsilon\psi, \qquad \text{(A.6)}$$

where the first and second terms represent kinetic and potential energy *operators*, respectively, and are expressed in terms of momentum, $\hat{p}=-i\hbar\, d/dq$, and position $\hat{q}$, operators, with ω defined by, $\omega=\sqrt{k/m}$. As *conjugate* operators, $\hat{p}$ and $\hat{q}$ obey a *commutation* relation, namely, $[\hat{q},\hat{p}]=\hat{q}\hat{p}-\hat{p}\hat{q}=i\hbar$, which indicates that the order in which they are applied is important. $\hbar$ is Planck's constant $(6.626\times10^{-34}\,J-\text{sec})$ divided by 2π. Furthermore, as conjugate operators, they also obey an uncertainty relation, namely, $\Delta\hat{q}\Delta\hat{p}\geq\hbar$, which gives the uncertainty in their values. A state prepared such that, say, $(\Delta\hat{q})^2<\hbar/2$, is called a *squeezed* state. Such a state lowers the uncertainty in one operator at the expense of that in the other [183].

To repeat ourselves, solving Schrödinger's operator equation, $\hat{H}|\psi>=\varepsilon|\psi>$, entails finding the *eigenvalues*, ε, giving the possible energies (frequencies) of the particle, and their corresponding *eigenvectors*, ψ, giving the wavefunctions that describe propagation in the system. For example, when the particle in question refers to atoms, separated by a distance *a*, undergoing longitudinal vibration modes in a monatomic linear chain (MLC), described by the Hamiltonian,

$$H_{MLC} = \sum_n \left(\frac{1}{2m} \hat{p}_n{}^2 + \frac{m\omega^2 \hat{q}_n{}^2}{2} \right), \tag{A.7}$$

then the eigenvalues (frequency dispersion curve) are, $\omega_q = \sqrt{\frac{4k}{m}} \left| \sin\left(\frac{qa}{2} \right) \right|$, and the eigenfunctions (propagating modes), $q_n = \xi \cdot \exp(i(qna - \omega t))$ [64]. Since, comparing (A.6) and (A.7), it is obvious that the latter is the sum of the Hamiltonian of *n* "particles," the MLC may be visualized as consisting of a set of *n* particles vibrating independently. In the context of the MLC, in which the vibrations represent acoustic waves, such fictitious particles are, in fact, called *phonons*, and (A.7) implies that the *state* of the MLC, in particular, its total energy, may be specified by giving the number *n* of "particles" present.

A.4 Creation and Annihilation Operators

It turns out that making the association:

$$a_n e^{-i\omega_n t} = \frac{1}{\sqrt{2m_n \hbar \omega_n}} \left(m_n \omega_n \hat{q}_n + i\hat{p}_n \right), \tag{A.8}$$

$$a_n^+ e^{i\omega_n t} = \frac{1}{\sqrt{2m_n \hbar \omega_n}} \left(m_n \omega_n \hat{q}_n - i\hat{p}_n \right), \tag{A.9}$$

where a_n and a_n^+ are new operators obeying the commutation relations $\left[a_n, a_{n'}^+\right] = \delta_{nn'}$, and $\left[a_n, a_{n'}\right] = \left[a_n^+, a_{n'}^+\right] = 0$, the *quantized* Hamiltonian (A.7) may be written as,

$$\hat{H}_{MLC} = \hbar \sum_n \omega_n \left(a_n^+ a_n + \frac{1}{2} \right). \tag{A.10}$$

Then, using the Hamiltonian expressed as in (A.11), and the commutation relations for the new operators, it can be shown that following result is follows,

$$\hat{H}_{MLC} \left| n \right> = \hbar \sum_n \omega_n \left(a_n^+ a_n + \frac{1}{2} \right) = \hbar \sum_n \omega_n \left(n + \frac{1}{2} \right) \left| n \right> = E_{MLC} \left| n \right>. \tag{A.11}$$

This means that if the field contains *n* phonons, the result of measuring its energy gives,

$$E_{MLC} = \hbar \sum_n \omega_n \left(n + \frac{1}{2} \right). \tag{A.12}$$

However, if the field contains no phonons (n=0), the energy is *not zero*, but is given by,

$$E_{MLC} = \frac{1}{2} \hbar \sum_n \omega_n . \tag{A.13}$$

This, *n=0,* state is called the *vacuum* state, and the corresponding energy, is *called zero-point energy*. Notice that, since $n = 0,\ 1,\ 2,\ 3 \ldots \infty$, the zero-point vacuum energy is, in principle, *infinite!* In practice, however, various factors, such as, dielectric constant cutoff, preclude it from becoming infinity, although still very large.

It we imagine the free-space in which a z-directed, x-polarized electromagnetic wave propagates as being divided into cubes of volume $V = L^3$, then, the solution to its associated electric field wave equation,

$$\nabla^2 E_x = \frac{1}{c^2} \frac{\partial^2 E_x}{\partial t^2}, \tag{A.14}$$

may be obtained by separation of variables as,

$$E_x(z,t) = \sum_n a_n q_n(t) f_n(z), \tag{A.15}$$

where, subject to the spatial boundary conditions $E_x(z=0,t) = E_x(z=L,t) = 0$, one obtains,

$$\left[\frac{d^2}{dz^2} + k_n^2 \right] f_n(z) = 0 \rightarrow f_n(z) = \sin(k_n z), \tag{A.16}$$

where,

$$k_n = n\pi / L \ , n = 1,\ 2,\ 3 \ldots \infty, \tag{A.17}$$

and,

$$\left[\frac{d^2}{dt^2} + \omega_n^2 \right] q_n(t) = 0 \rightarrow q_n(t) = \cos(\omega_n t - \phi), \tag{A.18}$$

where $\omega_n = ck_n$. Writing the electric field solution as

$$E_x(z,t) = \sum_n a_n q_n(t) \sin(k_n z), \tag{A.19}$$

the magnetic field is immediately obtained from Maxwell's equation, $\nabla \times E = -\partial B/\partial t$, as,

$$H_y(z,t) = \sum_n a_n \left(\frac{\dot{q}_n(t)\varepsilon_0}{k_n} \right) \cos(k_n z), \tag{A.20}$$

and the total field energy (Hamiltonian), which is given by,

$$H_{Field} = \frac{1}{2} \int_V \left(\varepsilon_0 E_x^2 + \mu_0 H_y^2 \right) dV , \tag{A.21}$$

becomes, upon substituting (A.19) and (A.20) into (A.21),

$$H_{Field} = \frac{1}{2} \sum_n \left(m_n \omega_n^2 q_n^2 + m_n \dot{q}_n^2 \right) = \frac{1}{2} \sum_n \left(m_n \omega_n^2 q_n^2 + \frac{p_n^2}{m_n} \right), \tag{A.22}$$

provided one makes the associations: $m_n = a_n^2 V \varepsilon_0 / 2\omega_n^2$, and $p_n = m_n \dot{q}_n$. The fact that each term in (A.22) is identical to the energy of a harmonic oscillator of frequency ω_n, implies that the field may be visualized as consisting of (or being populated by) a number n of such oscillators (*photons*), and the analysis given above follows directly. Accordingly, we can write

$$E_{Field} = \hbar \sum_n \omega_n \left(n + \frac{1}{2} \right). \tag{A.23}$$

Again, ideally for $n=0$, it is concluded that the electromagnetic vacuum possesses infinite energy. Furthermore, it can be shown [183] that the averages of the field and its magnitude squared are $\overline{E}_x = 0$ and $\overline{E}_x^{\,2} = 2|E_x|^2 (n + 1/2)$, where $E_n = \sqrt{\hbar\omega_n / \varepsilon_0 V}$ has dimensions of electric field. Thus, even when there is no field present, n=0, the vacuum is endowed with a *non-zero root-mean-square* deviation. These are called *zero-point vacuum fluctuations* and are the essence of the Casimir effect [19].

A.5 Second Quantization [232], [233]

Systems like the monatomic linear chain and the electromagnetic field, whose behavior can be described in terms of *fictitious* particles, such as phonons and photons, respectively, permeate many branches of physics, in particular, condensed matter physics. In the most general case, when described in terms of these *discrete* particles, the system is said to be represented in the second quantization or number representation formalism. The term *second-quantization* derives from the fact that in this theory the stuff the systems are made of, via this representation in terms of discrete particles, become quantized, i.e., an aggregate of discrete particles. You will recall that in the *first quantization,* it was the motion of the particles that became quantized. A second-quantized system may exhibit particle creation and annihilation, and multi-body interactions, and the formalism of second quantization (or number representation) has been devised to deal with the complex dynamics of these systems, in particular, for keeping track of the large number, and the statistics, of the particles that may be involved. The formalism, thus, prescribes ways to succintly represent pertinent wavefunctions and operators. The mathematical space in which second-quantized operators and vectors reside is called *Fock* space.

The simplest case occurs when the system has only one particle in, say, the state α, where this state is completely specified by giving pertinent quantum numbers, e.g., particle momentum, spin and spin projection. In this case, the one-particle state is represented by the ket $|1_\alpha\rangle$, and is taken as produced by the operation of the creation operator a_α^+ on the vacuum state $|0_\alpha\rangle$, the state of the system when there are no particles present. Mathematically, this is expressed by,

$$|1_\alpha\rangle = a_\alpha^+|0_\alpha\rangle . \tag{A.24}$$

If the system can contain many *noninteracting* particles, where the state of each particle, say, α, β, γ, δ, etc., respectively, is described by its respective set of quantum numbers, then the state representation in the second quantization formalism would be given by,

$$|1_\alpha\rangle = a_\alpha^+|0_\alpha,0_\beta,0_\gamma,0_\delta,\ldots\rangle = |1_\alpha,0_\beta,0_\gamma,0_\delta,\ldots\rangle . \tag{A.25}$$

If the system is in the vacuum state, i.e., there are no particles present in state α, then its state is represented by the ket $|0_\alpha\rangle$, which is taken as produced by the operation of the annihilation operator a_α on occupied single-particle state $|1_\alpha\rangle$. Mathematically, this is expressed as,

$$|0_\alpha\rangle = a_\alpha|1_\alpha\rangle . \tag{A.26}$$

In general, creation and annihilation operators are associated with each specific particle. Thus, it would be imposible to annihilate a particle in the state β with the annihilation operator for state α, i.e.,

$$a_\alpha \left|1_\beta\right\rangle = 0, \quad \beta \neq \alpha . \tag{A.27}$$

Since, using α to label a general state, $\left|1_\alpha\right\rangle = a_\alpha^+ \left|0_\alpha\right\rangle$, one can express (A.27) as,

$$a_\alpha \left|1_\beta\right\rangle = a_\alpha a_\beta^+ \left|0_\beta\right\rangle = \delta_{\alpha\beta} \left|0_\beta\right\rangle . \tag{A.28}$$

When the system contains many particles in multiple states, say, three particles in state γ, and one particle in state δ, following the above, the state may be represented by,

$$\left|3_\gamma 1_\delta\right\rangle = a_\gamma^+ a_\gamma^+ a_\gamma^+ a_\delta^+ \left|0\right\rangle , \tag{A.29}$$

where $\left|0\right\rangle$ represents the vacuum state.

The particles involved in second quantization may be identical or distinct. Due to the specificity/correspondence of the creation and annihilation operators with the state on which they operate, for any two single-particle states α and β, describing the system, the states,

$$\left|1_\alpha 1_\beta\right\rangle \sim a_\alpha^+ a_\beta^+ \left|0\right\rangle ; \; \left|1_\beta 1_\alpha\right\rangle \sim a_\beta^+ a_\alpha^+ \left|0\right\rangle , \tag{A.30}$$

must be identical, except for a freely chosen phase factor. If the phase factor is taken as real, then equating the two expressions gives,

$$\left|1_\alpha 1_\beta\right\rangle = \left|1_\beta 1_\alpha\right\rangle \Rightarrow a_\alpha^+ a_\beta^+ = a_\beta^+ a_\alpha^+ . \tag{A.31}$$

From knowldege that two identical boson are described by a symmetric wavefunction, it is deduced that this expression gives the commutation relation for bosons. On the other hand, from knowledge that two identical fermions are described by an antisymmetric wavefunction, it is deduced that

$$\left|1_\alpha 1_\beta\right\rangle = -\left|1_\beta 1_\alpha\right\rangle \Rightarrow a_\alpha^+ a_\beta^+ = -a_\beta^+ a_\alpha^+ , \tag{A.32}$$

gives the anticommutation relation for fermions. The anticommutation relations for fermions embody the fact that two fermions cannot occupy the same state (they obey Pauli's exclusion principle), i.e., $a_\alpha^+ a_\alpha^+ = -a_\alpha^+ a_\alpha^+ = 0$.

Other commutation relations are obtained as follows. For bosons, taking the hermitian conjugate of (A.31) yields,

$$a_\alpha a_\beta = a_\beta a_\alpha . \tag{A.33}$$

In addition, the facts that $a_\alpha|0\rangle = 0$ and $a_\alpha a_\beta^+|0\rangle = \delta_{\alpha\beta}|0\rangle$, mean that we can write $a_\alpha a_\beta^+|0\rangle - a_\beta^+ a_\alpha|0\rangle = \delta_{\alpha\beta}|0\rangle$, since the second term is zero, so that we also have the commutation relation,

$$a_\alpha a_\beta^+ - a_\beta^+ a_\alpha = \delta_{\alpha\beta} . \tag{A.34}$$

The operator whose eigenvalue measures the number of particles in a given state, say, state α, is the *number operator*, given by,

$$N_\alpha = a_\alpha^+ a_\alpha , \tag{A.35}$$

whereas the total number of particles, including all the distinct states, is given by,

$$N_{Total} = \sum_\alpha N_\alpha . \tag{A.36}$$

To measure (count) the number of particles in a given state, the operator N is applied to that state's eigenvector. The eigenvector of a state populated by n particles is described by the application of the creation operator n times, i.e.,

$$\left(a^+\right)^n|0\rangle = \underbrace{a^+a^+ \ldots a^+}_{n}|0\rangle , \tag{A.37}$$

so, measuring its occupation is effected by,

$$N\left(a^+\right)^n|0\rangle = N\underbrace{a^+a^+ \ldots a^+}_{n}|0\rangle = a^+a \cdot \underbrace{a^+a^+ \ldots a^+}_{n}|0\rangle . \tag{A.38}$$

Now, using the commutation relation (A.34), we can substitute $aa^+ \rightarrow 1 + a^+a$ in (A.38), so it becomes,

$$\begin{aligned} N\left(a^+\right)^n|0\rangle &= N\underbrace{a^+a^+ \ldots a^+}_{n}|0\rangle = a^+(1 + a^+a) \cdot \underbrace{a^+a^+ \ldots a^+}_{n-1}|0\rangle \\ &= a^+a^+(2 + N)\underbrace{a^+a^+ \ldots a^+}_{n-2}|0\rangle = \ldots \\ &= \underbrace{a^+a^+ \ldots a^+}_{n}(n + N)|0\rangle \\ &= \left(a^+\right)^n(n + a^+a)|0\rangle = n\left(a^+\right)^n|0\rangle \end{aligned} \tag{A.39}$$

So, the eigenvalue of N is n. For fermion operators, the pertinent anticommutation relations and number operator are,

$$a_\alpha a_\beta + a_\beta a_\alpha = \{a_\alpha, a_\beta\} = 0, \ \{a_\alpha^+, a_\beta^+\} = 0, \ \{a_\alpha, a_\beta^+\} = 0, \tag{A.40a}$$

$$a_\alpha^+ a_\alpha^+ = 0, \ a_\alpha a_\alpha = 0, \ N_\alpha = a_\alpha^+ a_\alpha, \tag{A.40b}$$

$$N_\alpha^2 = a_\alpha^+ a_\alpha a_\alpha^+ a_\alpha = a_\alpha^+ (1 - a_\alpha^+ a_\alpha) a_\alpha = a_\alpha^+ a_\alpha = N_\alpha . \tag{A.41}$$

The second quantization formalism is completed by the necessary expressions for operators, which upon acting on the wavefunction will measure certain quantities of interest. In this context, the quantities to be measured are classified according to the number of fundamental particles producing it. For instance, in a noninteracting system, these quantities may depend on individual particles, where each particle contributes its share independently from the others. An example of such a quantity is the kinetic energy of the system. On the other hand, quantities such as the Coulomb interaction energy, in an interacting system, depend on two-particle potentials, thus two-particle operators must be employed. Next, expressions for one- and two-particle operators are presented.

A typical one-particle operator is the kinetic energy. For a bosonic system, this is obtained by counting the number of particles in a given state, multiplying this number by the energy of each particle, and then adding the energies of all states. If an arbitrary particle occupies state α, following this prescription, then the one-particle operator is given by,

$$\hat{H} = \sum_\alpha \langle \alpha | K | \alpha \rangle N_\alpha = \sum_\alpha \langle \alpha | K | \alpha \rangle a_\alpha^+ a_\alpha . \tag{A.42}$$

In this expression, $\langle \alpha | K | \alpha \rangle$ is the state energy, given by,

$$E_\alpha = \langle \alpha | K | \alpha \rangle = \int \phi_\alpha^+(\vec{r}) \left[-\frac{\hbar^2}{2m} \nabla^2 \right] \phi_\alpha(\vec{r}) d\vec{r} , \tag{A.43}$$

where m is the particle mass, and $\phi_\alpha(\vec{r})$ is the configuration space representation of the wavefunction. Notice, that $\langle \alpha | K | \alpha \rangle$ may be computed in any convenient basis in which the wavefunction are available. Thus, in momentum-space basis we would have,

$$E_\alpha = \langle \alpha | K | \alpha \rangle = \int \phi_\alpha^+(\vec{p}) \left[\frac{p^2}{2m} \right] \phi_\alpha(\vec{p}) \frac{d\vec{p}}{(2\pi)^3} . \tag{A.44}$$

In the most general case, the one-particle energy operator is given by,

$$\hat{H} = \sum_{\alpha,\beta} \langle \alpha | K | \beta \rangle a_\alpha^+ a_\beta . \tag{A.45}$$

The kinetic energy of a one-particle state, is given by,

$$\begin{aligned} \langle 1_\rho | \hat{H} | 1_\sigma \rangle &= \langle 0 | a_\rho \hat{H} a_\sigma^+ | 0 \rangle = \sum_{\alpha\beta} \langle \alpha | K | \beta \rangle \langle 0 | a_\rho a_\alpha^+ a_\beta a_\sigma^+ | 0 \rangle \\ &= \sum_{\alpha\beta} \langle \alpha | K | \beta \rangle \delta_{\alpha\rho} \delta_{\beta\sigma} = \langle \rho | K | \sigma \rangle \end{aligned} . \tag{A.46}$$

where use was made of the identity $a_\beta a_\sigma^+ = \delta_{\beta\sigma}$. In the case of the two-particle potential,

$$V = \sum_{i<j} v(i,j) = \frac{1}{2} \sum_{i \neq j} v(i,j) , \tag{A.46}$$

the second-quantization operator, is given by,

$$V = \frac{1}{2} \sum_{\alpha\beta\gamma\delta} \langle \alpha\beta | v | \gamma\delta \rangle a_\alpha^+ a_\beta^+ a_\delta a_\gamma , \tag{A.47}$$

where the two-particle interaction energy may be evaluated in any basis. One typical source of confusion in this equation is the nature of the order of the annihilation elements in the number operator, in particular, the fact that instead of having $a_\alpha^+ a_\beta^+ a_\gamma a_\delta$, we have $a_\alpha^+ a_\beta^+ a_\delta a_\gamma$. This is done to make the expression valid for both bosons, where $a_\gamma a_\delta = a_\delta a_\gamma$, and fermions, where $a_\gamma a_\delta = -a_\delta a_\gamma$. Thus, for fermions the concomitant sign reversal will be automatically present. The matrix element, in configuration space, is given similarly as for the one-particle case, namely,

$$\langle \alpha\beta | v | \gamma\delta \rangle = \int \phi_\alpha^+(\vec{r}) \phi_\beta^+(\vec{r}) v(\vec{r}, \vec{r}') \phi_\gamma(\vec{r}) \phi_\delta(\vec{r}) d\vec{r} d\vec{r}' . \tag{A.48}$$

The two-particle interaction energy is given in terms of the wave functions as,

$$\begin{aligned} \langle 1_m 1_n | V | 1_p 1_q \rangle = \frac{1}{2} \int\int & \left[\phi_m^+(x_1) \phi_n^+(x_2) - \phi_n^+(x_1) \phi_m^+(x_2) \right] \\ & \times v(x_1, x_2) \left[\phi_p(x_1) \phi_q(x_2) - \phi_q(x_1) \phi_p(x_2) \right] dx_1 dx_2 \end{aligned} , \tag{A.49a}$$

and may be expressed in second-quantization noation as,

$$
\begin{aligned}
\langle 1_\rho 1_\sigma | V | 1_\tau 1_\nu \rangle &= \frac{1}{2}\sum_{\alpha\beta\gamma\delta} \langle \alpha\beta | v | \gamma\delta \rangle \langle 0 | a_\sigma a_\rho a_\alpha^+ a_\beta^+ a_\delta a_\gamma a_\tau^+ a_\nu^+ | 0 \rangle \\
&= \frac{1}{2}\sum_{\alpha\beta\gamma\delta} \langle \alpha\beta | v | \gamma\delta \rangle \left[\left(\delta_{\alpha\rho}\delta_{\beta\sigma} \pm \delta_{\alpha\sigma}\delta_{\beta\rho} \right) \left(\delta_{\alpha\rho}\delta_{\beta\sigma} \pm \delta_{\alpha\sigma}\delta_{\beta\rho} \right) \right] \\
&= \frac{1}{2}\left[\langle \rho\sigma | v | \tau\nu \rangle + \langle \sigma\rho | v | \nu\tau \rangle \pm \left(\langle \sigma\rho | v | \tau\nu \rangle + \langle \rho\sigma | v | \nu\tau \rangle \right) \right] \\
&= \langle \rho\sigma | v | \tau\nu \rangle \pm \langle \rho\sigma | v | \nu\tau \rangle
\end{aligned} \quad . \text{(A.49b)}
$$

A.5.1 Field Operators

A common practice in the application of the number representation formalism in interacting (many-body) systems is to express the Hamiltonians in terms of so-called *field operators*, $\psi(x)$ and $\psi^+(x)$, which are defined by,

$$\psi(x) = \sum_i \phi_i(x) c_i \, , \quad \text{(A.50)}$$

and

$$\psi^+(x) = \sum_i \phi^*{}_i(x) c_i^+ \, . \quad \text{(A.51)}$$

The field operators obey the commutation relations,

$$
\begin{aligned}
\{\psi(x), \psi(x')\} &= \sum_{i,j} \phi_i(x)\phi_j(x') c_i c_j + \sum_{i,j} \phi_j(x')\phi_i(x) c_j c_i \\
&= \sum_{i,j} \phi_i(x)\phi_j(x') \{c_i, c_j\} = 0
\end{aligned} \quad , \quad \text{(A.52)}
$$

$$
\begin{aligned}
\{\psi(x), \psi^+(x')\} &= \sum_{i,j} \phi_i(x)\phi_j^*(x') c_i c_j^+ + \sum_{i,j} \phi_j^*(x')\phi_i(x) c_j^* c_i \\
&= \sum_{i,j} \phi_i(x)\phi_j^*(x') \{c_i, c_j^+\} = 0 \\
&= \sum_i \phi_i(x)\phi_i^*(x') = \delta(x - x')
\end{aligned} \quad , \quad \text{(A.53)}
$$

where the latter simply expresses the completeness relation of the wavefunctions.

For example, in terms of the field operators, the one-particle Hamiltonian operator,

$$\hat{H}_0 = \sum_{\alpha,\beta} \langle \alpha | K | \beta \rangle c_\alpha^+ c_\beta, \tag{A.54}$$

where,

$$\langle \alpha | K | \beta \rangle = \int \phi_\alpha^*(x) K(x) \phi_\beta(x) dx, \tag{A.55}$$

is expressed as,

$$\hat{H}_0 = \int \psi^+(x) K(x) \psi(x) dx. \tag{A.56}$$

This is proven by substituting (A.53) and (A.54) into (A.59) to recover (A.57):

$$\begin{aligned} \int \sum_\alpha \phi_\alpha^*(x) c_\alpha^+ K(x) \sum_\beta \phi_\beta(x) c_\beta dx &= \sum_{\alpha,\beta} c_\alpha^+ c_\beta \int \phi_\alpha^*(x) K(x) \phi_\beta(x) dx \\ &= \sum_{\alpha,\beta} \langle \alpha | K | \beta \rangle c_\alpha^+ c_\beta \end{aligned}. \tag{A.57}$$

An interpretation of the field operators is obtained by operating with them on the vacuum state. For inatsnce, operating with $\psi^+(x)$ on $|0\rangle$, we obtain,

$$\begin{aligned} \psi^+(x)|0\rangle &= \sum_\alpha \phi_\alpha^+(x) c_\alpha^+ |0\rangle \\ &= \sum_\alpha \phi_\alpha^+(x) \phi_\alpha(x_\alpha) = \delta(x - x_\alpha) \end{aligned}, \tag{A.58}$$

since the operation of the creation operator of the state α on the vacuum creates a particle there. This results indicates that $\psi^+(x)$ behaves as the creator or a particle at position x. Similarly, one obtains that $\psi(x)$ destroys a particle at position x.

In the context of this book, the second quantization formalism is key to the presentation on the Luttinger liquid. This deals with the description of electrons constrained to move in one dimension and described by the Hamiltonian,

$$\hat{H} = \hat{H}_0 + \hat{H}_{Int}, \tag{A.59}$$

where the first term gives the electron kinetic energy,

$$\hat{H}_0 = \sum_{\sigma=\uparrow,\downarrow} \sum_k E_k c_{k\sigma}^+ c_{k\sigma} , \tag{A.60}$$

and the second term gives the electron-electron Coulomb interaction,

$$\hat{H}_{Int} = \frac{1}{2L} \sum V_{kk'q\sigma\sigma'}^{\sigma\sigma'} c_{k+q\sigma}^+ c_{k'-q\sigma'}^+ c_{k'\sigma'} c_{k\sigma} . \tag{A.61}$$

Solving for the eigenvalues and eigenfunctions of the problem is facilitated by modeling the fermions in terms of bosons, in which case the Hamiltonian becomes diagonal and it is easy to solve. The procedure that accomplishes this fermion-to-boson transformation is called *bosonization*, and is presented Appendix B.

Appendix B

BOSONIZATION

B.1 Introduction

The method of bosonization consists in modeling a fermionic system by an equivalent bosonic system, with the advantage that the diagonalization of the bosonized Hamiltonian of the fermionic system becomes easier [138]. This fact becomes more transparent upon comparing the 1D specific heats for a solid with sound velocity c_s, obtained by Debye c_L^{Debye}, and that for a Fermi gas of noninteracting electrons with Fermi velocity v_F, obtained by Pauli c_L^{Pauli},

$$c_L^{Debye} = \frac{\pi}{3} k_B \left(\frac{k_B T}{\hbar c_s} \right), \qquad \text{(B.1a)}$$

$$c_L^{Pauli} = \frac{\pi}{3} k_B \left(\frac{k_B T}{\hbar v_F} \right). \qquad \text{(B.1b)}$$

Clearly, replacing $c_s \Leftrightarrow v_F$ one obtains identical results.

B.2 Bosonization "Rules"

While many works attempting to explain bosonization have been published, a particularly lucid and very pedagogical treatment was that advanced by Delft and Schoeller [139]. They clearly expose, in a systematic fashion the procedure of bosonization, and we follow their exposition closely.

In general, bosonizing a theory involving M species of fermions may be accomplished when a fix specific set of conditions are met, in particular:

1) The theory can be formulated in terms of a set of fermion creation and annihilation operators, c^+_{kq} and c^+_{kq}, which obey the following canonical anti-commutation relations

$$\left\{c_{k\eta}, c^+_{k\eta}\right\} = \delta_{\eta\eta'}\delta_{kk'}, \tag{B.2}$$

where the index $\eta = 1,...M$ labels M different species, which might be present, and the index $k \in [-\infty, \infty]$ is a discrete, unbounded wave number of the form,

$$k = \frac{2\pi}{L}\left(n_k - \frac{1}{2}\delta_b\right), \quad \text{with } n_k \in Z \text{ and } \delta_b \in [0,2), \tag{B.3}$$

with n_k are integers, L is a length associated with the size of the system, and δ_b is a parameter that embodies the nature of the boundary conditions of the fermion fields, i.e., whether they periodic or fixed. According to Delft and Scholler [139], in typical examples η can denote electron spin: η = (↑, ↓), in which case M = 2, or it distinguishes left-moving from right-moving spinless electrons, as found in a one-dimensional wire, in which case: η = (L,R), and M = 2. k refers to the momentum index that labels the energy states, E_k, of a free noninteracting Fermi gas, defined with respect to the Fermi energy, so that $E_0 = E_F$. The discrete and unbounded nature of k are needed in order to enable the systematic accounting of the states, on the one hand, and the proper definition of bosonic operators, oh the other.

2) The fermion fields are defined in terms of the creation and annihilation operators as follows,

$$\psi_\eta(x) = \sqrt{\frac{2\pi}{L}} \sum_{k=-\infty}^{\infty} e^{-ikx} c_{k\eta}, \tag{B.4}$$

with inverse,

$$c_{k\eta} = \frac{1}{\sqrt{2\pi L}} \int_{-L/2}^{L/2} e^{ikx} \psi_\eta(x) dx, \tag{B.5}$$

where $x \in [-L/2, L/2]$, but may be allowed to go to infinity (L $\to \infty$), at the conclusion of the procedure, if necessary. The fields $\psi_\eta(x)$ and the variable x are, in general, mathematical constructs which result from the development undertaken to formulate the model in terms of the operators $c_{k\eta}$. In particular, for discrete *k*, $\psi_\eta(x)$ obeys the following properties:

$$\psi_\eta(x + L/2) = e^{i\pi\delta_b}\psi_\eta(x - L/2), \tag{B.6}$$

where $\delta_b = 0$ for the periodicity condition and $\delta_b = 1$ for anti-periodicity.

3) The fermionic number representation (Fock) space is reorganized so that the Fock space of states spanned by the operators $c_{k\eta}$ is rendered as a direct sum, $F = \sum_{\oplus \vec{N}} H_{\vec{N}}$ over the Hilbert spaces $H_{\vec{N}}$ characterized by a fixed particle number $\vec{N}$, within each of which all excitations are *bosonic*, i.e., particle-hole-like. The first step towards accomplishing this is to define the vacuum state $|0\rangle$ by,

$$c_{k\eta}|0\rangle \equiv 0 \text{ for } k > 0, \quad (n_k > 0), \tag{B.7}$$

$$c_{k\eta}^{+}|0\rangle \equiv 0 \text{ for } k \le 0, \quad (n_k \le 0). \tag{B.8}$$

(B.7) signifies that states above k=0 are empty, therefore, none may be destroyed, and states below k=0 are all occupied, therefore, none may be populated. The occupation of all other states in Fock space are defined relative to the vacuum, particularly the operation of fermion normal ordering with respect to it. A function is said to be in fermion-normal-order form when all $c_{k\eta}$ with k>0, and all $c_{k\eta}^{+}$ with $k \le 0$ are positioned to the right of all other operators $c_{k\eta}^{+}$ with k>0 and $c_{k\eta}$ with $k \le 0$. Thus, for operators $A, B, C, \ldots \in \{c_{k\eta}; c_{k\eta}^{+}\}$, this is represented by,

$$\,^{+}_{+}ABC\ldots^{+}_{+} = ABC\ldots - \langle 0|ABC\ldots|0\rangle. \tag{B.9}$$

4) The number operator $\hat{N}_\eta$ possesses eigenvalues $\vec{N} = (N_1, N_2, \ldots, N_M) \in Z^M$, whose aggregate makes up the $\vec{N}$-particle Hilbert space $H_{\vec{N}}$. In particular, $\hat{N}_\eta$ counts the number of electrons of species η, is given by,

$$\hat{N}_\eta \equiv \sum_{k=-\infty}^{\infty} {}^{+}_{+}c^{+}_{k\eta}c_{k\eta}{}^{+}_{+} = \sum_{k=-\infty}^{\infty}\left[c^{+}_{k\eta}c_{k\eta} - \langle 0|c^{+}_{k\eta}c_{k\eta}|0\rangle\right], \tag{B.10}$$

and operates on states of the form $\left|\vec{N}\right\rangle$. The ground state, denoted $\left|\vec{N}\right\rangle_0$, represents the state in which there are no particle-hole excitations, and it is constructed as follows.

$$\left|\vec{N}\right\rangle_0 \equiv (C_1)^{N_1}(C_2)^{N_2}...(C_M)^{N_M}|0\rangle, \tag{B.11}$$

where,

$$(C_\eta)^{N_\eta} \equiv \begin{cases} c^{+}_{N_\eta\eta}c^{+}_{(N_\eta-1)\eta}...c^{+}_{1\eta} & \text{for} \quad N_\eta > 0, \\ 1 \quad \text{for} \quad N_\eta = 0, & \\ c^{+}_{(N_\eta+1)\eta}c^{+}_{(N_\eta+2)\eta}...c^{+}_{0\eta} & \text{for} \quad N_\eta < 0. \end{cases}, \tag{B.12}$$

5) Given the fixed number of particles in every $\vec{N}$ -particle Hilbert space, their excitations are construed as particle-hole excitations of the ground state $\left|\vec{N}\right\rangle_0$, and captured by bosonic creation and annihilation operators defined by,

$$b^{+}_{q\eta} = \frac{i}{\sqrt{n_q}}\sum_{k=-\infty}^{\infty} c^{+}_{k+q\eta}c_{k\eta}, \qquad b_{q\eta} = \frac{-i}{\sqrt{n_q}}\sum_{k=-\infty}^{\infty} c^{+}_{k-q\eta}c_{k\eta}, \tag{B.13}$$

where $q = \dfrac{2\pi n_q}{L} > 0$, and $n_q \in Z^+$ is a positive integer. Then, operating on any state $\left|\vec{N}\right\rangle$ with $b^{+}_{q\eta}$ or $b_{q\eta}$ causes an aggregate of particle-hole excitations, where each excitation's momentum differs, from that in the ground state, by q units. This permits interpreting $b^{+}_{q\eta}$ and $b_{q\eta}$ as momentum raising and lowering operators, which obey the following commutation relations:

$$\left[b_{q\eta}, b_{q'\eta'}\right] = \left[b^{+}_{q\eta}, b^{+}_{q'\eta'}\right] = 0, \ \left[N_{q\eta}, b_{q'\eta'}\right] = \left[N_{q\eta}, b^{+}_{q'\eta'}\right] = 0, \text{ for all } q, q', \eta, \eta' \tag{B.14}$$

$$\left[b_{q\eta}, b^{+}_{q'\eta'}\right] = \delta_{\eta\eta'}\delta_{qq'} \tag{B.15}$$

6) The bosonic vacuum states, the ground states given by $\left|\vec{N}\right\rangle_0$, are defined such that,

$$b_{q\eta}\left|\vec{N}\right\rangle_0 = 0, \quad for \quad all\, q,\eta \tag{B.16}$$

and admit a boson-normal-ordering protocol, in which all $b_{q\eta}$ are moved to the right of all $b^{+}_{q\eta}$, so that, for operators $A, B, C, \ldots \in \left\{b_{k\eta}, b^{+}_{k\eta}\right\}$, this is represented by,

$$\overset{+}{+} ABC\ldots \overset{+}{+} = ABC\ldots - \left\langle \vec{N}_0 \middle| ABC\ldots \middle| \vec{N}_0 \right\rangle. \tag{B.17}$$

7) Every state $\left|\vec{N}\right\rangle$ in the $\vec{N}$-particle Hilbert space, may be generated by acting on the ground state $\left|\vec{N}\right\rangle_0$ by a properly chosen bilinear combination of the fermion operators, $\left|\vec{N}\right\rangle = \bar{f}\left(c^{+}_{k\eta}c_{k\eta}\right)\left|\vec{N}\right\rangle_0$, or of boson operators, $\left|\vec{N}\right\rangle = f\left(b^{+}\right)\left|\vec{N}\right\rangle_0$.

8) There exist raising and lowering (ladder) operators whose action on a given state $\left|\vec{N}\right\rangle$ of the $\vec{N}$-particle Hilbert space changes the total number of fermions by one. These operators are called *Klein factors*, denoted F^{+}_{η} and F_{η}, respectively, and obey the following properties, namely,

a) They commute with all bosonic operators, i.e.,

$$\left[b_{q\eta}, F_{\eta'}\right] = \left[b^{+}_{q\eta}, F^{+}_{\eta'}\right] = \left[b_{q\eta}, F_{q'\eta'}\right] = \left[b^{+}_{q\eta}, F_{\eta'}\right] = 0, \ \text{for all} \ q,\eta,\eta' \tag{B.18}$$

b) Their action on a state $\left|\vec{N}\right\rangle$ of the $\vec{N}$-particle Hilbert space may be expressed as the product of a particle-hole excitations $f\left(b^{+}\right)$ acting on the corresponding $\vec{N}$-particle ground state $\left|\vec{N}\right\rangle_0$, in particular,

$$F^{+}_{\eta}\left|\vec{N}\right\rangle \equiv f\left(b^{+}\right)c^{+}_{N\eta+1}\left| N_1,\ldots N_\eta,\ldots N_M\right\rangle_0 \equiv f\left(b^{+}\right)\hat{T}_{\eta}\left| N_1,\ldots N_\eta+1,\ldots N_M\right\rangle_0 \tag{B.19}$$

$$F_\eta \left| \vec{N} \right\rangle \equiv f(b^+) c_{N\eta} \left| N_1,...N_\eta,...N_M \right\rangle_0 \equiv f(b^+) \hat{T}_\eta \left| N_1,...N_\eta - 1,...N_M \right\rangle_0 \quad \text{(B.20)}$$

where, $\hat{T}_\eta$ is the so-called the *phase-counting operator*,

$$\hat{T}_\eta \equiv (-1)^{\sum_{\eta=1}^{\eta-1} \hat{N}_\eta} \quad \text{(B.21)}$$

which keeps track of the number of signs picked up when acting with a fermion operator $c_{k\eta}$ on a state $\left| \vec{N} \right\rangle_0$ to produce a different state $\left| \vec{N}' \right\rangle_0$, i.e.,

$$c_{k\eta} (C_1)^{N_1} ..(C_\eta)^{N_\eta} ...(C_M)^{N_M} |0\rangle = \hat{T}_\eta (C_1)^{N_1} ..(C_{\eta-1})^{N_{\eta-1}} c_{k\eta} (C_\eta)^{N_\eta} ..(C_M)^{N_M} |0\rangle . \quad \text{(B.22)}$$

c) The Klein factors obey commutation relations,

$$\{F_\eta^+, F_{\eta'}\} = 2\delta_{\eta\eta'} \quad \text{for all} \quad \eta, \eta', \quad \text{(B.23)}$$

$$F_\eta F_{\eta'}^+ = F_\eta^+ F_\eta = 1, \quad \text{(B.24)}$$

$$\{F_\eta^+, F_\eta^+\} = \{F_\eta, F_\eta\} = 0, \quad \text{for} \quad \eta \neq \eta', \quad \text{(B.25)}$$

$$\left[\hat{N}_\eta, F_{\eta'}^+\right] = \delta_{\eta\eta'} F_\eta^+, \quad \left[\hat{N}_\eta, F_{\eta'}\right] = -\delta_{\eta\eta'} F_\eta . \quad \text{(B.26)}$$

B.3 Bosonic Field Operators

In analogy with fermion field operators, boson fields operators, $\phi_\eta(x)$, are defined in terms of bosonic operators as follows:

$$\varphi_\eta(x) \equiv -\sum_{q>0} \frac{1}{\sqrt{n_q}} e^{-iqx} b_{q\eta} e^{-aq/2}, \text{ and } \varphi_\eta^+(x) \equiv -\sum_{q>0} \frac{1}{\sqrt{n_q}} e^{iqx} b_{q\eta}^+ e^{-aq/2}, \quad \text{(B.27)}$$

with

$$\phi_\eta(x) \equiv \varphi_\eta(x) + \varphi_\eta^+(x) = -\sum_{q>0} \frac{1}{\sqrt{n_q}} \left(e^{-iqx} b_{q\eta} + e^{iqx} b_{q\eta}^+ \right) e^{-aq/2}, \quad \text{(B.28)}$$

where $\varphi_\eta(x)$ and $\phi_\eta(x)$ are constructed to exhibit periodicity L in x, and $a > 0$ is an artifact to regularize the divergent momentum sums $(q \to \infty)$, with its reciprocal $1/a$ interpreted as the maximum momentum difference for the $c^+_{k\pm q} c_k$-combinations occurring in the fermionic functions ϕ. The bosonic fields obey the commutation relations,

$$\left[\varphi_\eta(x), \varphi_{\eta'}(x')\right] = \left[\varphi^+_\eta(x), \varphi^+_{\eta'}(x')\right] = 0\,, \tag{B.29}$$

$$\begin{aligned}\left[\varphi_\eta(x), \varphi^+_{\eta'}(x')\right] &= \delta_{\eta\eta'} \sum_{q>0} \frac{1}{n_q} e^{-q[i(x-x')+a]} \\ &= \delta_{\eta\eta'} \ln\left[1 - e^{-i\frac{2\pi}{L}(x-x'-ia)}\right] \\ &\xrightarrow{L\to\infty} \delta_{\eta\eta'} \ln\left[i\frac{2\pi}{L}(x-x'-ia)\right]\end{aligned}, \tag{B.30}$$

where use was made of the identity $\ln(1-y) = -\sum_{n=1} y^n / n$. In terms of these bosonic fields the normal-ordered electron density becomes a function of $\partial_x \phi_\eta(x)$, as follows,

$$\begin{aligned}\rho_\eta(x) &\equiv {}^+_+\psi^+_\eta(x)\psi_\eta(x)^+_+ / 2\pi = \frac{1}{L}\sum_q e^{-iqx} \sum_k {}^+_+ c^+_{k-q,\eta} c_{k\eta}{}^+_+ \\ &= \frac{1}{L}\sum_{q>0} i\sqrt{n_q}\left(e^{-iqx} b_{q\eta} - e^{iqx} b^+_{q\eta}\right) + \sum_k {}^+_+ c^+_{k-q,\eta} c_{k\eta}{}^+_+ \,. \\ &= \frac{\partial_x \phi_\eta(x)}{2\pi} + \frac{1}{L}\hat{N}_\eta \quad \text{for } (a \to 0)\end{aligned} \tag{B.31}$$

B.4 Bosonization Identity and Its Application to Hamiltonian with Linear Dispersion

The ultimate purpose of the preliminaries presented thus far, has been to enable familiarization with the mathematical language and techniques required to effect the transformation of Hamiltonians expressed in terms of fermionic field operators, into Hamiltonians expressed in terms of the bosonic field operators. This transformation is enabled by the *bosonization identity*,

$$\psi_\eta(x) = F_\eta a^{-1/a} e^{-i\frac{2\pi}{L}\left(\hat{N}_\eta - \frac{1}{2}\delta_b\right)x} e^{-i\phi_\eta(x)} \xrightarrow{L\to\infty} F_\eta a^{-1/a} e^{-i\phi_\eta(x)}. \quad \text{(B.32)}$$

The derivation of this identity was undertaken by Delft and Schoeller [138] in two steps. First, the demonstration that $\psi_\eta(x)|N\rangle_0$ is an eigenstate of the bosonic operator $b_{q\eta}$ was undertaken, which guarantees that $\psi_\eta(x)|N\rangle_0$ may be expressed as a *coherent state*, and then the consequences of acting with ψ_η on a general state were determined.

The relationship between ψ_η and $b_{q\eta}$ is captured by their commutation relation which, in turn, derives from their respective definitions given in (B.5) and (B.8). The pertinent commutation relations are,

$$\left[b_{q\eta'}, \psi_\eta(x)\right] = \delta_{\eta\eta'}\alpha_q\psi_\eta(x), \quad \text{(B.33)}$$

$$\left[b^+_{q\eta'}, \psi_\eta(x)\right] = \delta_{\eta\eta'}\alpha^*_q\psi_\eta(x), \quad \text{(B.34)}$$

where $\alpha_q(x) = \frac{i}{\sqrt{n_q}} e^{iqx}$. Applying (A.91) on the ground state, we obtain,

$$b_{q\eta'}, \psi_\eta(x)\left|\vec{N}\right\rangle_0 - \psi_\eta(x) b_{q\eta'}\left|\vec{N}\right\rangle_0 = \delta_{\eta\eta'}\alpha_q\psi_\eta(x)\left|\vec{N}\right\rangle_0. \quad \text{(B.35)}$$

However, since $b_{q\eta}\left|\vec{N}\right\rangle_0 = 0$, the second term vanishes and we get the result,

$$b_{q\eta}\psi_\eta(x)\left|\vec{N}\right\rangle_0 = \alpha_q\psi_\eta(x)\left|\vec{N}\right\rangle_0, \quad \text{(B.36)}$$

which shows that $\psi_\eta(x)\left|\vec{N}\right\rangle_0$ is an eigenvector of $b_{q\eta}$, the boson annihilation operator, with eigenvalue α_q.

A well known result of quantum mechanics is that if a state is an eigenvector of the annihilation operator, then this state is a *coherent state* [139]. A coherent state has many useful properties. For instance, its uncertainty relation is minimized, i.e., $\Delta x\Delta p = \hbar/2$. Such a state may be expressed in the form,

$$\psi_\eta(x)\left|\vec{N}\right\rangle_0 = \exp\left[\sum_{q>0}\alpha_q(x) b^+_{q\eta}\right] F_\eta\hat{\lambda}_\eta(x)\left|\vec{N}\right\rangle_0, \quad \text{(B.37)}$$

where $\hat{\lambda}_\eta$ is a phase operator, and F_η effects the η-particle removal normally effected by $\psi_\eta(x)$. Inserting (C.27) into (B.37), the following expression is obtained,

$$\psi_\eta(x)\left|\vec{N}\right\rangle_0 = e^{-i\varphi_\eta^+(x)} F_\eta \hat{\lambda}_\eta(x)\left|\vec{N}\right\rangle_0 . \tag{B.38}$$

Use of the operator identity, $e^{-B}Ae^{B} = A + C$ or $\left[A, e^{B}\right]Ce^{B}$, and identification of $A = b_{q\eta'}$, $B = -i\varphi_\eta^+(x)$, and $C = \delta_{\eta\eta'}\alpha_q(x)$, secures the compliance of (B.38) with (B.36).

The crux of the bozonization identity lies on (B.38). According to Delft and Schoeller [139], this expression embodies the fact that acting with the fermionic field $\psi_\eta(x)$ on $\left|N\right\rangle_0$ may effect the removal of one η-particle from the ground state in two ways. First, via the interpretation of $\psi_\eta(x)$ as $\sqrt{\frac{2\pi}{L}}\sum_{k=-\infty}^{\infty} e^{-ikx} c_{k\eta}$, it creates an infinite linear combination of *single-hole states* caused by each application of the fermion annihilation operator $c_{k\eta}$, see Fig. B-1.

$$\sum_{n=0}^{\infty} y^n c_{-n}\left|0\right\rangle_0 = \left[\quad + \; y \quad + \; y^2 \quad + \ldots \right]$$

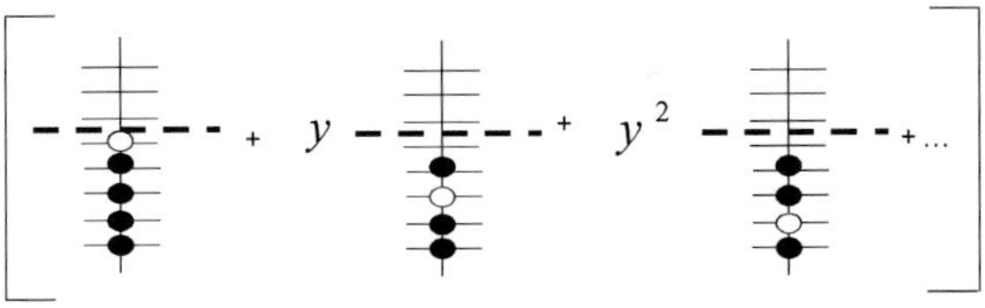

Figure B-1. Effect of acting with $\psi_\eta(x)$ on the ground state. We have expressed $\psi(x) \sim \sum_{n=0}^{\infty} y^n c_{-n}$, with $y = e^{i2\pi x/L}$. (*After* [139].)

On the other hand, observing the right-hand side of (B.38), this η-particle removal may also be effected by removing the highest η-electron from $\left|N\right\rangle_0$, which yields a different ground state, namely, $c_{N_\eta\eta}\left|\vec{N}\right\rangle_0$, and then creation of a linear combination of hole states through the action of the boson creation operators $b_{q\eta}^+$ present in $e^{-i\varphi_\eta^+(x)}$. The effect of first operating with the Klein factor is shown in Fig. B-2, and that of operating with the field operator is shown in Fig. B-3..

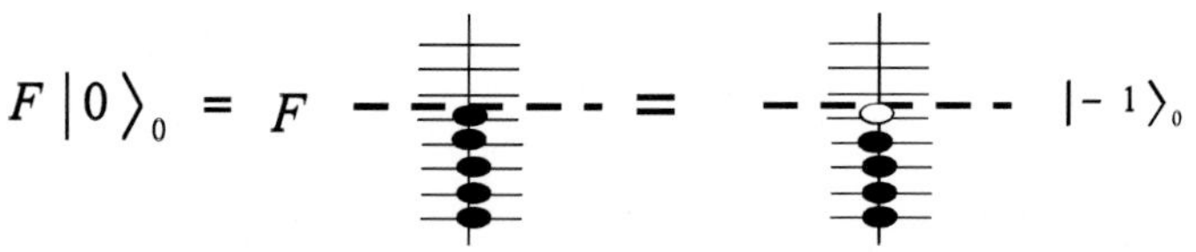

Figure B-2. Effect of acting on the ground state by the Klein factor. All levels are move down by one, thus creating a hole at the top level.

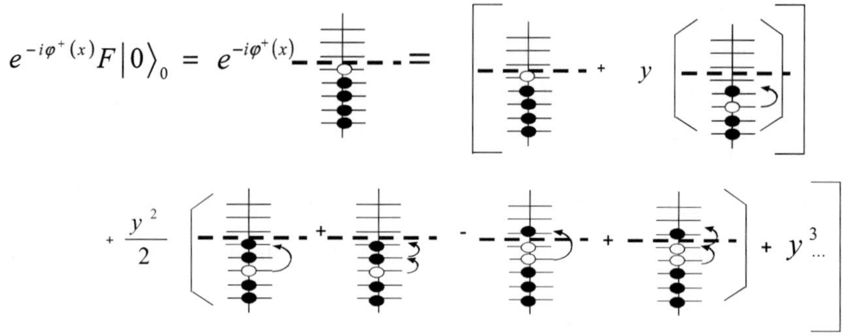

Figure B-3. Effect of acting with $\psi_\eta(x)$ on the ground state. We have expressed $\psi(x) \sim e^{-i\varphi^+(x)}F$. (*After* [139].)

The value of the operator $\hat{\lambda}_\eta(x)$ is determined by Delft and Schoeller [139] to be,

$$\hat{\lambda}_\eta(x) = \sqrt{\frac{2\pi}{L}}\, e^{-i\left(\hat{N}_\eta - \frac{1}{2}\delta_b\right)x}. \tag{B.39}$$

It may be shown [138], by example, that Figs. B-1 and B-2 are equivalent, i.e., that,

$$\sqrt{\frac{2\pi}{L}} \sum_{n\in Z} e^{-i\left(n-\frac{1}{2}\delta_b\right)2\pi x/L} c_n|0\rangle_0 = F_\eta \sqrt{\frac{2\pi}{L}} \sum_{n\in Z} e^{-i\left(N_\eta-\frac{1}{2}\delta_b\right)2\pi/L} e^{-i\varphi_\eta^+(x)}|0\rangle_0 . \tag{B.40}$$

This involves writing,

$$\sum_{n=0}^{\infty} y^n c_{-n}\left|0\right\rangle_0 = e^{-\left(\sum_{n=1}^{\infty}\frac{1}{n} y^n \rho_n\right)} c_0\left|0\right\rangle_0, \text{ where } y \equiv e^{i2\pi x/L}, \text{ and } \rho_n \equiv \sum_{\bar{n}\in Z} c^+_{\bar{n}+n} c_n$$

$$= \left[1 - y\rho_1 + y^2\left(-\frac{1}{2}\rho_2 + \frac{1}{2}\rho_1^2\right) + y^3\left(\frac{1}{3}\rho_3 + \frac{1}{2}\rho_1\rho_2 - \frac{1}{6}\rho_1^3\right) + \ldots\right] c_0\left|0\right\rangle_0 \quad . \quad \text{(B.41)}$$

$$= \sum\left[A_n y^n c_{-n} + B_n y^{n+2}\left(c^+_{n+1}c_{-1}\right)c_0 + C_n y^{n+3}\left(c^+_{n+1}c_{-2}\right)c_0 + \ldots\right]\left|0\right\rangle_0$$

From (B.41) it is observed that the only nonzero coefficients in it are $A_n = 1$, whereas all others, namely, $B_n = C_n = \ldots = 0$. This signifies that whenever $e^{-i\varphi^+(x)}$ acts on $c_0\left|0\right\rangle_0$, all the possible ways in which states of the form $c_n^+ c_{-n} c_0\left|0\right\rangle_0$ may be excited interfere destructively, so that only terms of the form $y^n c_{-n}\left|0\right\rangle_0$ interfere constructively. This can be seen by considering (B.41) after inserting the sums ρ_n. In that case for the A_n coefficient one obtains,

$$\sum_{n=0}^{\infty} A_n y^n c_{-n}\left|0\right\rangle_0 = \left\{1 - y\left(c_0^+ c_{-1}\right) + y^2\left[-\frac{1}{2}\left(c_0^+ c_{-2}\right) + \frac{1}{2}\left(c_{-1}^+ c_{-2}\right)\left(c_0^+ c_{-1}\right)\right]\right.$$
$$\left. + y^3\left[-\frac{1}{3}\left(c_0^+ c_{-3}\right) + \frac{1}{2}\left(c_{-2}^+ c_{-3}\right)\left(c_0^+ c_{-2}\right) - \frac{1}{6}\left(c_{-2}^+ c_{-3}\right)\left(c_{-1}^+ c_{-2}\right)\left(c_0^+ c_{-1}\right)\right] + \ldots\right\} c_0\left|0\right\rangle_0 , \quad \text{(B.42)}$$
$$= \left\{1 + y c_{-1} + y^2\left[\frac{1}{2} + \frac{1}{2}\right] c_{-2} + y^3\left[\frac{1}{3} + \frac{1}{2} + \frac{1}{6}\right] c_{-3} + \ldots\right\}\left|0\right\rangle_0$$

whereas for the B_n one obtains,

$$\sum_{n=0}^{\infty} B_n y^{n+2}\left(c_n^+ c_{-1}\right)c_0\left|0\right\rangle_0 = \left\{y^2\left[-\frac{1}{2}\left(c_1^+ c_{-1}\right) + \frac{1}{2}\left(c_1^+ c_0\right)\left(c_0^+ c_{-1}\right)\right]\right.$$
$$\left. + y^3\left[-\frac{1}{3}\left(c_2^+ c_{-1}\right) + \frac{1}{2}\left(c_2^+ c_1\right)\left(c_1^+ c_{-1}\right) - \frac{1}{6}\left(c_2^+ c_1\right)\left(c_1^+ c_0\right)\left(c_0^+ c_{-1}\right)\right] + \ldots\right\} c_0\left|0\right\rangle_0 \quad . \quad \text{(B.43)}$$
$$= \left\{y^2\left[-\frac{1}{2} + \frac{1}{2}\right]\left(c_1^+ c_{-1}\right) + y^3\left[-\frac{1}{3} + \frac{1}{2} - \frac{1}{6}\right]\left(c_2^+ c_{-1}\right) + \ldots\right\}\left|0\right\rangle_0$$

Clearly, all the $A_n = 1$ and $B_n = 0$. These examples together with Figs. B-1 and B-2 should provide an intuitive way of assimilating the concept of bosonization. What we will do next is to finally present an example of the bosonization procedure, namely, their Delft and Scholler's [139] application of the procedure to a Hamiltonian with a linear dispersion.

They begin by assuming a linear dispersion of the form, $E(k) = v_F \hbar k$, which measures all energies in units $v_F \hbar$, where the total Hamiltonian is,

$$H_0 \equiv \sum_{\eta} H_{0\eta}, \tag{B.44}$$

with,

$$\begin{aligned} H_{0\eta} &\equiv \sum_{k=-\infty}^{\infty} {}^{+}_{+} k c^{+}_{k\eta} c_{k\eta} {}^{+}_{+} \\ &= \int_{-L/2}^{L/2} \frac{dx}{2\pi} {}^{+}_{+} \psi^{+}_{\eta}(x) i\partial_x \psi_{\eta}(x) {}^{+}_{+} \end{aligned} . \tag{B.45}$$

Then, the fact that the Hamiltonian commutes with the number operator $\left[H_{0\eta}, \hat{N}_{\eta'}\right] = 0$ for all η, η', is exploited as an argument to justify that any $\vec{N}$-particle ground state is an eigenstate of $H_{0\eta}$, in particular, $H_{0\eta}\left|\vec{N}\right\rangle_0 = E^{\vec{N}}_{0\eta}\left|\vec{N}\right\rangle_0$. The eigenvalue is obtained by adding the energy of all levels,

$$\begin{aligned} E^{\vec{N}}_{0\eta} = {}_0\left\langle \vec{N}\right| H_{0\eta} \left|\vec{N}\right\rangle_0 &= \frac{2\pi}{L} \begin{cases} \sum_{n=1}^{N_\eta} (n - \delta_b/2) = \frac{1}{2} N_\eta^2 + \frac{1}{2} N_\eta (1-\delta_b) & for\ N_\eta \geq 0, \\ \sum_{n=N_\eta+1}^{0} -(n-\delta_b/2) = \frac{1}{2} N_\eta^2 + \frac{1}{2} \left|N_\eta\right| (1-\delta_b) & for\ N_\eta < 0 \cdot \end{cases} \\ &= \frac{2\pi}{L} \frac{1}{2} N_\eta (N_\eta + 1 - \delta_b) \end{aligned} \tag{B.46}$$

This gives the ground state energy of $H_{0\eta}$. When the system is excited, its eigenstate energy $\left|E\right\rangle$, may increase in units of q. This follows from the commutation relations,

$$\left[H_{0\eta}, b^{+}_{q\eta'}\right] = q b^{+}_{q\eta'} \delta_{\eta\eta'}, \tag{B.47}$$

and its consequence,

$$H_{0\eta} b_{q\eta}^{+} \left| E \right\rangle = (E+q) b_{q\eta}^{+}, \tag{B.48}$$

and implies that the $b_{q\eta}^{+}$ acting on the ground state $\left| \vec{N} \right\rangle_0$ may generate the complete $\vec{N}$-particle Hilbert space. The bosonic variables then may be employed to represent $H_{0\eta}$, including both the ground and excited states. This is accomplished when it takes the bosonized form,

$$H_{0\eta} = \sum_{q>0} q b_{q\eta}^{+} b_{q\eta} + \frac{2\pi}{L}\frac{1}{2}\hat{N}_\eta \left(\hat{N}_\eta + 1 - \delta_b\right). \tag{B.49}$$

Since $H_{0\eta}$ does not change the particle number, no Klein factors, F_η^{+}, appear.

B.5 Bosonization Treatment of Spinless Electrons in One-Dimensional Wire

The one-dimensional wire is the prototypical system of a Lüttinger liquid. It is described as a one-dimensional conductor of length L with free spinless left- and right-moving electrons. The electrons possess momentum $p \in (-\infty, \infty)$, and propagate according to a dispersion relation given by $E(p) = \left(p^2 - p_F^2\right)/2m$. Since electrons are confined to move either to the left or to the right in a 1D conductor, the usual fermion field,

$$\Psi_{phys}(x) \equiv \sqrt{\frac{2\pi}{L}} \sum_{p=-\infty}^{\infty} e^{ipx} c_p, \tag{B.50}$$

is expressed as,

$$\Psi_{phys}(x) = \sqrt{\frac{2\pi}{L}} \sum_{k=-k_F}^{\infty} \left(e^{-i(k_F+k)x} c_{-k_F-k} + e^{i(k_F+k)x} c_{k_F+k} \right), \tag{B.51}$$

where the momentum p is written as $p = \mp(k + k_F)$, with $k \in \left[-k_{F,} \infty\right)$, and $p < 0$ corresponds to the left(L)-moving electrons, and $p > 0$ to the right(R)-moving electrons. In the context of our species definitions, the index $\nu = (L, R)$ plays the analogous role to η.

We now effect the bosonization procedure described previously. First, one must make $k \in \left[-k_{F,} \infty\right)$ unbounded from below and discrete. This is accomplished by artificially extending the range of k to be unbounded, introducing L- and R-moving fermion fields $\widetilde{\psi}_{L/R}$, and imposing boundary conditions (B.Cs) on these to discretize k. Making $k \in (-\infty, \infty)$ entails defining energies of the form $E_{k,v} \equiv E(0) + v_F(k + k_F)$ in the range $k < -k_F$. These additional "unphysical"states do not alter the low-energy physics of the system, however, a strong perturbation, such as might be due to an electric field or an impurity, then the procedure would not apply because of the larger energies involved [139]. Extending the range of k, the fermionic field Ψ_{phys} is written in terms of fields representing L- and R-moving electrons which now possess the unbounded k define above. This new fermionic field takes the form,

$$\Psi_{phys}(x) = e^{-ik_F x}\widetilde{\psi}_L(x) + e^{+ik_F x}\widetilde{\psi}_R(x), \tag{B.52}$$

where,

$$\widetilde{\psi}_{L/R}(x) = \sqrt{\frac{2\pi}{L}} \sum_{k=-\infty}^{\infty} e^{\mp ikx} c_{k,L/R} \,. \tag{B.53}$$

Lastly, imposing B.C.s quantizes the fermion fields momentum. If these are taken as anti-periodic, we have, $\widetilde{\psi}_{L/R}(L/2) = -\widetilde{\psi}_{L/R}(-L/2)$, which implies $\delta_b = 1$. Having defined the prerequisite conditions for bosonization, the consequent number operators, Klein factors, and boson operators, $\hat{N}_{L/R}$, $F_{L/R}$, and $b_{qL/R}$ are defined in terms of the fermion annihilation operator $c_{kL/R}$. This results in the following,

$$\widetilde{\phi}_{L/R}(x) = -\sum_{n_q \in Z^+} \frac{1}{\sqrt{\mathrm{n_q}}} e^{-aq/2}\left[e^{\mp iqx} b_{qL/R} + e^{\mp iqx} b^+_{qL/R}\right] \left(q = \frac{2\pi}{L} n_q > 0\right), \tag{B.54}$$

$$\widetilde{\psi}_{L/R}(x) \equiv a^{-1/2} F_{L/R} e^{\mp i\frac{2\pi}{L}\left(\hat{N}_{L/R} - \frac{1}{2}\delta_b\right)x} e^{-i\widetilde{\phi}_{L/R}(x)}, \tag{B.55}$$

$$\widetilde{\rho}_{L/R}(x) \equiv {}^+_+ \widetilde{\psi}^+_{L/R} \widetilde{\psi}_{L/R}{}^+_+ = \pm\partial_x \widetilde{\phi}_{L/R}(x) + \frac{2\pi}{L}\hat{N}_{L/R}, \tag{B.56}$$

where the boundary conditions $\widetilde{\phi}_{L/R}(L/2) = \widetilde{\phi}_{L/R}(-L/2)$ (periodic) on the bosons and density fields have been imposed. Notice that, while the density $\widetilde{\rho}_{L/R}$ is quadratic in the fermion field, it is only linear in the boson field. This is key to the simplification brought about by the bosonization procedure.

REFERENCES

[1] R. P. Feynman, "There's plenty of room at the bottom," *J. Microelectromech. Syst.*, vol. 2, Mar. 1992, pp. 60-66.

[2] G.E. Moore, "Cramming more components onto integrated circuits," *Electronics*, vol. 38, Number 8, April 19, 1965.

[3] L. Esaki, "New phenomenon in narrow Germanium pn-junctions," *Phys. Rev.*, vol. 109, 1957, pp. 603-604.

[4] T. A. Fulton and G. J. Dolan, "Observation of single-electron charging effects in small tunnel junctions," *Phys. Rev. Lett.,* vol. 59, 1987, pp.109–112.

[5] L.L. Chang, L. Esaki, and R. Tsu, "Resonant tunneling in semiconductor double-barriers," *Appl. Phys. Lett.,* vol. 24, 1974, p. 593.

[6] B J van Wees *et al.* "Quantised conductance of point contacts in a two dimensional electron gas," *Phys. Rev. Lett.* , vol. 60, 1988, pp. 848-850.

[7] *Proc. Int. Symp. On Nanostructure Physics and Fabrication*, College Station, TX, March 13-15, M.A. Reed and W.P. Kirk, Editors., Academic Press, San Diego (1989).

[8] *Proc. Int. Symp. On Nanostructure Physics and Fabrication*, Santa Fe, NM, May 20-24, 1991, W.P. Kirk and W.P. Kirk, Editors., Academic Press, San Diego (1992).

[9] H. Kroemer, Theory of a wide-gap emitter for transistors," Proc. IRE, vol. 45, 1957, pp. 1535-1537.

[10] W.P. Dumke, J.M. Woodall and. V.L. Rideout, "GaAs-GaAlAs heterojunction transistor for high frequency operation," *Solid State Electron.*, vol. 15, 1972, pp. 1339-1344.

[11] P. M. Asbeck, D.L. Miller, W.C. Petersen, *et al*, "GaAs/GaAlAs heterojunction bipolar transistors with cutoff frequencies above 10 GHz," *IEEE Electron Device Letts.*, vol. EDL-3, 1982, pp. 366-368.

[12] S.L. Su, et al., "Double heterojunction AlGaAs/GaAs bipolar transistors by MBE with a current gain of 1650," *IEEE Electron Device Letts.*, vol. EDL-4, 1983, pp. 130-132.

[13] R.J. Malik, J.R. Hayes, F. Capasso et al., "High-gain AlInAs/GaInAs transistors grown

by molecular beam epitaxy*," IEEE Electron Device Letts.*, vol. EDL-4, 1982, pp. 366-368.

[14] T. Miura, S. Hiyamizuk, T. Fujii *et al.*, "A new field effect transistor woth selectively doped GaAs/n-AlGaAs heterostructures," *Jpn. J. Appl. Phys*, vol. 19, 1980, pp. L255-L277.

[15] A. B. Frazier, A. B., R.O. Warrington, *et al.,* "The Miniaturization Technologies: Past, Present, and Future," *IEEE Trans. Ind. Electronics*, vol. 42, 1995, p.423.

[16] G. Binnig and H. Rohrer, "Scanning tunneling microscopy," Helvetica Physica Acta 55, 1982, pp. 726–735.

[17] P. J. F. Harris, *Carbon Nanotubes and Related Structures*. Cambridge Univ. Press, Cambridge, U.K. (1999).

[18] H.J. De Los Santos, *Introduction to Microelectromechanical (MEM) Microwave Systems*, Norwood, MA: Artech House (1999).

[19] H. B. G. Casimir, "On the attraction between two perfectly conducting plates," *Proc. K. Ned.. Akad. Wet.*, vol. 51, 1948, pp. 793–799.

[20] P.W. Milonni, *The Quantum Vacuum: An Introduction to Quantum Electrodynamics*, Academic Press: San Diego (1994).

[21] C. Itzykson and J.-B. Zuber, *Quantum Field Theory* (McGraw-Hill International Editions: New York, N.Y. (1985).

[22] H.J. De Los Santos, "Nanoelectromechanical Quantum Circuits and Systems," *Proc. IEEE,* vol. 91, No. 11, November 2003, pp. 1907-1921.

[23] S.M. Sze, *VLSI Technology*, Editor, New York: McGraw-Hill (1983).

[24] Jaeger, R. C., *Introduction to Microelectronics Fabrication*, Volume V, Modular Series on Solid State Devices, G.W. Neudeck and R.F. Pierret, Eds., Addison-Wesley (1988).

[25] Hiroshi Toshiyoshi, "Micro-Opto-Electro Mechanical Systems (MOEMS) and Their Applications," Short Course Notes.

[26] S. Wolf and R. N, Tauber*, Silicon Processing for the VLSI Era: Process Technology*, Lattice Press (1986)

[27] H.J. De Los Santos, *Introduction to Microelectromechanical (MEM) Microwave Systems*, Second Edition, Norwood, MA: Artech House (2004).

[28] C. Kittel, *Introduction to Solid-State Physics*, New York, NY: John Wiley & Sons, Inc. (1986).

[29] K.R. Williams and R.S. Muller, "Etch Rates for Micromachining Processing," *JMEMS*, vol. 5, No. 4, Dec. 1996, pp. 256-269.

[30] G.T. A. Kovacs, N. I. Maluf, and K. E. Petersen, "Bulk Micromachining of Silicon," *Proc. IEEE*, vol. 86, No. 8, Aug. 1998, pp. 1536-1551.

[31] F. Laermer and A. Schlip, "Method of Anisotropically Etching Silicon," Patent # 5501893, Issued: March 26, 1996, Assignee: Robert Boch GmbH, Stuttgart, Federal Republic of Germany.

[32] F. Laermer, A. Schilp, K. Funk and M. Offenberg, "Bosch Deep Silicon Etching: Improving Uniformity and Etch Rate for Advanced MEMS Applications," 1999 IEEE

[33] K. A. Shaw, Z.L. Zhang and N.C. MacDonald, "SCREAM I: A single mask single-

crystal silicon process for microelectromechanical structures," *Sensons and Actuators* A 40 (1994), pp. 210-213.

[34] G. M. Whitesides and J. C. Love, "The Art of Building Small," *Scientific American*, Sept. 2001, pp. 39-47.

[35] S.M. Sze, *Physics of Semiconductor Devices*, New York: John Wiley & Sons (1981).

[36] F. Capasso and G. Margaritondo, Editors, *Heterojunction Band Discontinuities: Physics and Device Applications*, Amsterdam: North-Holland (1987).

[37] S.L. Chuang, *Physics of Optoelectronic Devices*, New York: John Wiley & Sons (1995).

[38] [Online]: http://www.veeco.com/learning/learning_molecularbeam.asp

[39] H.J. De Los Santos, *et al.* "Electron Transport Mechanisms in Abrupt- and Graded-Base/Collector AlInAs/GaInAs/InP DHBTs," *Int. Symp.Compound Semicond, Inst. Phys. Conf. Ser. No. 141*: Chapter 6, 1995, pp. 645-650.

[40] H.J. De Los Santos, *et al.*, "DHBT/RTD-Based Active Frequency Multiplier for Wireless Communications," *1997 IEEE Int. Symp. Compound Semicond*, Chapter 7, pp. 515-518, 1998.

[41] F. J. Giessibl, "Advances in atomic force microscopy," *Review of Modern Physics* ,vol. 75, 2003, p.949.

[42] A. A. Baski, "Fabrication of Nanoscale Structures using STM and AFM," *Advanced Semiconductor and Organic Nano-techniques*, Part 3 (edited by Morkoc)," Academic Press (2002)

[43] J. Israelachvili, Intermolecular and Surface Forces, 2nd ed. London: Academic Press, (1991).

[44] B.M. Law and F. Rieutord, "Electrostatic forces in atomic force microscopy", Phys. Rev. vol. B 66, 2002, pp. 035402–1–6.

[45] L. Olsson, N. Lin, V. Yakimov et al, "A method for in situ characterization of tip shape in ac-mode atomic force microscopy using electrostatic interaction", *J. Appl. Phys.*, vol. 84, 1998, pp. 4060–4064.

[46] P.J.F. Harris, *Carbon Nanotubes and Related Structures*, Cambridge University Press: Cambrudge, UK (1999).

[47] S. Li, Z Yu, G. Gadde et al., "Carbon Nanotube Growth for GHz Devices," *2003 NANO Conference*, August, San Francisco, CA.

[48] C. L. Cheung, J. H. Hafner, and C. M. Lieber, "Carbon nanotube atomic force microscopy tips: Direct growth by chemical vapor deposition and application to high-resolution imaging," *PNAS*, April 11, 2000, vol. 97 No. 8, p 3813.

[49] A. Requicha, " Nanorobots, NEMS and Nanoassembly," *Proc. IEEE,* vol. 91, No. 11, November 2003, pp. 1922-1933.

[50] S. Hong, J. Zhu, C.A. Mirkin, "Multiple Ink Nanolithography: Toward a Multiple-Pen Nano-Plotter," *Science* 283, 1999, p. 661.

[51] E. Yablonovitch, Inhibited Spontaneous Emission in Solid State Physics and Electronics," *Phys. Rev. Letts.*, Vol. 58, No. 20, 1987, pp. 2059-2062.

[52] H. J. De Los Santos, *Introduction to Microelectromechanical (MEM) Microwave Systems,* Artech House: Norwood, MA (1999).

[53] S.D. Senturia, *Microsystem Design,* Kluwer Academic Publishers: Boston, MA (2001).

[54] H. J. De Los Santos, *RF MEMS Circuit Design for Wireless Communications*, Artech House: Norwood, MA (2002).

[55] Online Version: http://www.intel.com/labs/BobRaoIDF022802.pdf

[56] H.J. De Los Santos, "Nanoelectromechanical Quantum Circuits and Systems," *Proc. IEEE,* Vol. 91, No. 11, November 2003, pp. 1907-1921.

[57] R. E. Collin, Foundations of Microwave Engineering, New York, N.Y.: *IEEE Press*, (2001).

[58] S. Datta, *Quantum Phenomena*, Reading, MA: Addison-Wesley (1989).

[59] T. Ando, Y. Arakawa, K. Furuya, *et al.* (Eds.), *Mesoscopic Physics and Electronics*, Springer-Verlag: New York (1999).

[60] C. Cohen-Tannoudji, B. Diu and F. Laloë, *Quantum Mechanics*, Volume 1, (John Wiley & Sons: New York (1977).

[61] Y.-Q. Li and B. Chen, "Quantum Theory of Mesoscopic Electric Circuits," *Phys. Rev. B* vol. 53, 1996, p. 4027.

[62] J.C. Flores, "Mesoscopic Circuits with Charge Discreteness: Quantum Transmission Lines," *Physical Review* B, vol. 64, Issue 23, December 15, 2001, p. 235309.

[63] W.H. Louisell, *Quantum Statistical Properties of Radiation*, John Wiley & Sons: New York (1973).

[64] N.W. Ashcroft and N.D. Mermin, *Solid State Physics*, Sounders College: Philadelphia, (1976).

[65] L. Lapidus and G.F. Pinder, *Numerical Solution of Partial Differential Equations in Science and Engineering*, John Wiley & Sons: New York (1982).

[66] J.C. Flores and E. Lazo, "Bloch-like oscillations induced by charge discreteness in quantum mesoscopic rings," Online Version: http://xxx.lanl.gov, cond-mat/9910101

[67] Y.-Q. Li and B. Chen, "Quantum Theory for Mesoscopic Electronic Circuits and Its Applications," OnlineVersion: http://xxx.lanl.gov, cond-mat/9907171

[68] G. Schön, "Single Electron Tunneling," Chapter 3 of the book on *Quantum Transport and Dissipation*, T. Dittrich, P. Hangge, G. Ingold, B. Kramer, G. Schon, W. Zwerger, (1997).

[69] K. K. Likharev, Single-Electron Devices and Their Applications," *Proc. IEEE*, vol. 87, April 1999, pp. 606-632.

[70] S. Babiker, "Simulation of Single-Electron Transport in Nanostructured Quantum Dots," *IEEE Trans. Electron Dev.*,vol. 52 , No. 3 , March 2005, pp. 392-396.

[71] S. M. Goodnick and J. Bird, "Quantum-Effect and Single-Electron Devices," *IEEE Trans. Nanotechnology*, vol. 2, No. 4, Dec. 2004, pp. 368-385.

[72] A.N. Cleland and M.L. Roukes, "Fabrication of high frequency nanometer scale mechanical resonators from bulk Si substrates", *Appl. Phys. Lett.*, vol. 69, 1996, p.2653.

[73] S. Sapmaz, Ya. M. Blanter, L. Gurevich, *et al*, "Carbon Nanotubes as Nanoelectromechanical Systems," *Phys. Rev*. B, vol. 67, 2003, pp. 235414-235422.

[74] H.B.G. Casimir, "On the attraction between two perfectly conducting plates," *Proc. K. Ned. Akad. Wet*. vol. 51, 1948, p. 793.

[75] P.W. Milonni, *The Quantum Vacuum: An Introduction to Quantum Electrodynamics*,

Academic Press: San Diego (1994).

[76] C. Itzykson and J.-B. Zuber, *Quantum Field Theory*, McGraw-Hill International Editions: New York, N.Y. (1985).

[77] L.S. Brown and G.J. Maclay, "Vacuum Stress between Conducting Plates: An Image Solution," *Phys. Rev.*, vol. 184, No. 5, August 1969, pp. 1272-1279.

[78] I.E. Dzyaloshinskii, "Condensed Matter Physics," Course, Dept. of Physics, UC Irvine, Spring 2002.

[79] M. Dequesnes, S.V. Rotkin, and N.R. Aluru, "Calculation of pull-in voltages for carbon-nanotube-based nanoelectromechanical switches," *Nanotechnology*, vol. 13, 2002, pp. 120-131.

[80] J. Maclay, P. Milonni, H. Fearn, "Of some theoretical signifiacnce: Implications of Casomir effects," *European Journal of Physics*, 2001.

[81] A. MacKinnon, "Quantum gears: a simple mechanical system in the quantum regime," Online Version: http://xxx.lanl.gov, cond-mat/0205647

[82] E. Buks and M.L. Roukes, "Stiction, adhesion energy, and the Casimir effect in micromechanical systems," *Phys. Rev* **B** vol. 63, No. 3, Jan. 15 2001.

[83] T.H. Boyer, "Van der Waals forces and zero-point energy for dielectric and permeable materials," *Phys. Rev.* A, vol. 9, No. 5, May 1974, pp. 2078-2084.

[84] O. Kenneth, I. Klich, A. Mann, and M. Revzen, "Repulsive Casimir forces," *Phys Rev Letters*, Vol. 89, No. 3, 15 July 2002. p. 0033001-1

[85] G. J. Maclay, "Unusual properties of conductive rectangular cavities in the zero point electromagnetic field: Resolving Forward's Casimir energy extraction cycle paradox," *PROCEEDINGS of STAIF-99* (*Space Technology and Applications International Forum-1999*, Albuquerque, NM, January, 1999), edited by M.S. El-Genk, AIP Conference Proceedings 458, American Institute of Physics, New York 1999. Online Version: http://www.quantumfields.com/articles.html

[86] G J. Maclay, R. Ilic, M. Serry, P. Neuzil, "Use of AFM (Atomic Force Microscope) Methods to Measure Variations in Vacuum Energy Density and Vacuum Forces in Microfabricated Structures," *NASA Breakthrough Propulsion Workshop*, Cleveland, Ohio, May, 1997. Online Version: http://www.quantumfields.com/articles.htm

[87] H. Bortman, "Energy Unlimited," *New Scientist Magazine*, Jan. 22, 2000, pp. 32-34.

[88] K A Milton, *The Casimir Effect: Physical Manifestations of Zero-point Energy* , World Scientific, Singapore (2001).

[89] M. Bordag, U. Mohideen, and V.M. Mostepanenko, "New Developments in the Casimir Effect," *Physics Reports* vol. 353, 2001, pp1-205.

[90] A. Roy and U. Mohideen, "A verification of quantum field theory—measurement of Casimir force," *Pramana—Journal of Physics*, Vol. 56, Nos 2 & 3, pp. 239-243.

[91] B.W. Harris, F. Chen, and U. Mohideen, "Precision measurement of the Casimir force using gold surfaces," *Physical Review A*, Nov. 2000, Vol.62, No.5, pp. 052109/1-5.

[92] S.K. Lamoreaux, "Experimental Verifications of the Casimir Attractive Force Between Solid Bodies," *Phys. Rev. Lett.* vol. 78, 1997, p. 5.

[93] S.K. Lamoreaux, "Demonstration of the Casimir Force in the 0.6 to 6 μm range, *Phys. Rev. Letts.*, vol. 78, No. 1, 1997, pp. 5-8.

[94] F. Chen , U. Mohideen, G.L. Klimchitskaya *et al.*, *"Demonstration of the lateral Casimir force," Phys. Rev. Lett.*, vol.88, No.10, 2002. pp.101801/1-4.

[95] G. Bressi, G. Carugno, R. Onofrio *et al.*, "Measurement of the Casimir force between parallel metallic surfaces," *Phys. Rev. Lett.* vol. 88, No. 4, 28 January 2002,

pp. 041804.
[96] J. Schwinger, L.L. De Radd, Jt., and K.A. Milton, "Casimir effect in dielectrics," *Ann. Phys. (NY)* vol. 115, 1978, p.1.
[97] V.B. Bezerra, G.L. Klimchitskaya, and C. Romero, Casimir force between a flat plate and a spherical lens: Application to the results of a new experiment," *Mod. Phys. Lett.*, **A12**, 1997, p. 2613.
[98] E.M. Lifshitz, The theory of molecular attractive forces between solids," *Sov. Phys. JETP* vol **2**, 1956, p. 73.
[99] J. Blocki, J. Randrup, W.J. Swiatecki and C.F. Tsang, "Proximity forces," *Ann. Phys.* (N.Y.) vol. 105, 1977, p. 427.
[100] T.P. Spiller, "Quantum Information Processing: Cryptography, Computation, and Teleportation," *Proc. IEEE*, vol. 84, No. 12, Dec. 1996, pp. 1719-1746.
[101] A. Barenco, "Quantum physics and computers," *Contemporary Physics*, vol. 37, 1996, pp.375-89
[102] A.M. Steane, "Quantum Computing", *Rept. Prog. Phys.* vol. 61, 1998, pp. 117-173.
[103] B. Schumacher, "Quantum coding," *Phys. Rev.* A 51, 1995, pp. 2738–2747.
[104] D. Deutsch, "Quantum theory, the Church-Turing principle and the universal quantum computer," *Proc. Roy. Soc. Lond.* A vol. 400, 1985, pp. 97-117.
[105] D. Deutsch, "Quantum computational networks," *Proc. Roy. Soc. Lond.* A vol. 425, 1989, pp. 73-90.
[106] C.H. Bennet, G. Brassard, C. Crepeau et al, "Teleporting an Unknown Quantum State via Dual Classical and Einstein-Podolsky-Rosen Channels," *Phys. Rev. Letts*, vol. 70, 29 March 1993, pp. 1895-1899.
[107] Kwiat, P., H. Weinfurter, T. Herzog et al. 1995, *Phys. Rev. Lett.* vol. 74, p. 4763.
[108] C.H. Bennett, "Quantum Information and Computation," *Physics Today*, Oct. 1995, pp. 24-30.
[109] V. Vedral and M.B. Plenio, "Basics of Quantum Computation," Prog. Quant. Electron vol. 22, 1998, pp.1-39.
[110] J. Preskill, "Quantum computingL pro and con," Proc. R. Soc. Lond. A, vol. 454, 1998, pp. 469-86.
[111] M. A. Nielsen and I. L. Chuang *Quantum Computation and Quantum Information*, Cambridge (2000).
[112] D. Vion, A. Aassime, A. Cottet et al., "Manipulating the Quantum State of an Electrical Circuit, *Science* vol. 296, 3 May 2002, pp. 886-889.
[113] R. Landauer, "Spatial Variation of Currents and Fields Due to Localized Scatterers in Metallic Conduction," *IBM J. Res. Dev.*, vol. 1, 1957, pp. 223-231.
[114] B. J. van Wees *et al.* "Quantised conductance of point contacts in a two dimensional electron gas," *Phys. Rev. Lett.* vol. 60, 1988, pp. 848-850.
[115] A. Szafer and A. D. Stone, "Theory of Quantum Conduction through a Constriction," *Phys. Rev. Letts.*, vol. 62, No. 3, 1989, pp. 300-303.
[116] L.L. Chang, L. Esaki, and R. Tsu, "Resonant tunneling in semiconductor double-barriers," Appl. Phys. Lett., Vol. 24, 1974, pp. 593.
[117] D.K. Roy, *Quantum Mechanical Tunnelling and Its Applications*, Singapore: World Scientific (1986).

[118] F. Capasso and G. Margaritondo (Editors), *Heterojunction Band Discontinuities: Physics and Device Applications*, Amsterdam: North-Holland (1987).

[119] H. Mizuta and T. Tanoue, *The Physics and Applications of Resonant Tunneling Diodes*, Cambridge, UK: Cambridge University Press (1995).

[120] D. H. Chow, J. N. Schulman, E. Ozbayet *et al.*, "1.7-ps microwave integrated circuit-compatible InAs/AlSb resonant tunneling diodes," *Appl. Phys. Lett.* vol. 61, 1992, pp. 1685-1687.

[121] D.D. Coon and H.C. Liu, "Tunneling currents and two-body effects in quantum well and superlattice structures," *App. Phys. Letts.*, vol. 47, No. 2, 1984, pp. 172-174.

[122] R. Lake and S. Datta, "Nonequilibrium Green's-function method applied to double-barrier resonant-tunneling diodes," *Phys. Rev.* B, vol. 45, No. 12, 1992, pp. 6670-6685.

[123] W.-R. Liou and P. Roblin, "High Frequency Simulation of Resonant Tunneling Diodes," *IEEE Trans. Electron Dev.*, Vol. 41, No. 7, 1994, pp. 1098-1111..

[124] J.N. Schulman, "$Ga_{1-x}Al_xAs$-$Ga_{1-y}Al_yAs$-GaAs double-barrier structures," *J. of Appl. Phys.* vol. 60, 1986, pp. 3954-3958.

[125] Y. Aharonov and D. Bohm, "Significance of Electromagnetic Potentials in Quantum Theory," *Phys. Rev.* Second Series, vol. 115, No. 3, 1959, pp. 485-491

[126] R.E. Peierls, *Quantum Theory of Solids*, Oxford: Clarendon Press (1955).

[127] L.G.C. Rego and G. Kirczenow, "Quantized thermal conductance of dielectric quantum wires," *Physical Review Letters*, Volume 81, Issue 1, July 6, 1998, pp. 232-235

[128] K. Schwab, E.A. Henriksen, J.M. Worlock et al., "Measurement of the Quantum of Thermal Conductance," *Nature*, vol. 404, 2002, pp.974-977.

[129] D.E. Angelescu, M.C. Cross, and M.L. Roukes, "Heat Transport in Mesoscopic Systems", *Superlattices and Microstructures* vol. 23, 1998, p.673.

[130] D.G. Cahill, W.K. Ford, K.E. Goodson *et al.*, "Nanoscale Thermal Transport," to appear in *Applied Phys. Rev., J. Appl. Phys.* vol. **93**, 2002, p. 1.

[131] A.A. Abrikosov, L.P. Gorkov, and I.E. Dzyaloshinskii, *Methods of Quantum Field Theory in Statistical Physics*," New York, N.Y.: Dover Publications (1963)

[132] S. Rodriguez, "Electron Theory of Solids," Course Notes, Dept. of Physics, Purdue University, 1987-88.

[133] H.J. Schulz, "Fermi liquids and non–Fermi liquids," Online Version: http:// xxx.lanl.gov cond-mat/9503150

[134] A. J. Schofield, "Non-Fermi liquids." *Contemporary Physics*, vol. 50, No. 2, 1999, pp. 95-115.

[135] R. Melin, B. Doucot, and P. Butaud, "Breakdown of the Fermi liquid picture in one dimensional fermion systems: connection with the energy level statistics," *Journal de Physique I*, vol. 4, No. 5, May 1994, p.737.

[136] M. Horsdal, *Interacting electrons in one dimension: The Luttinger model and the quantum Hall effect*, Ph.D. Dissertation, Department of Phyiscs,University of Oslo, May 2003.

[137] F.D.M. Haldane, "Properties of the Luttinger model and their extension to the general 1D interacting spinless Fermi Gas," *J. Phys.C: Solid State Phys.*, vol. 14, 1981, 2585.

[138] K. Schönhammer, "Lüttinger Liquids: The Basic Concepts," OnlineVersion: http://cond- mat/0305035

[139] J. von Delft and H. Schoeller, "Bosonization for Beginners — Refermionization for Experts," *Annalen der Physik*, vol. 4, 1998, pp. 225-305.

[140] C. L. Kane and M. P. A. Fisher, "Transport in a Single Channel Luttinger Liquid,"*Phys. Rev. Lett.* vol. 68, 1992, p. 1220.

[141] M.P.A. Fisher, L.I. Glazman. Transport in one-dimensional Luttinger liquid. In: *Mesoscopic Electron Transport,* ed. by L.L. Sohn, L.P. Kouwenhoven, and G. Schoen. NATO ASI Series, vol. 345, Kluwer Academic Publishers (1997). p. 331.

[142] S. John, "Strong Localization of Photons in Certain Disordered Dielectric Superlattices," *Phys. Rev. Letts*, vol. 58, No. 23, 8 June 1987, pp. 2486-2489.

[143] N. Hamada, S. Sawada and A. Oshiyama, "New one-dimensional conductors: graphitic microtubules," *Phys. Rev. Lett.*, vol. 68, 1992, p. 1579.

[144] P.L. McEuen, M.S. Fuhrer, and H. Park, "Single-Walled Carbon Nanotube Electronics," *IEEE Trans. Nanotechnology*, Vol. 1, No. 1, March 2002, pp. 78-85.

[145] P. Avouris, J. Appenzeller, R. Martel, and S.J. Wind, "Carbom Nanotube Electronics," *IEEE Proc.*, Vol. 91, No. 11, Nov. 2003, pp. 1772-1784.

[146] M.S. Dresselhaus, G. Dresselhaus and R. Saito, "Physics of carbon nanotubes," Carbon, vol. 33, 1995, p. 883.

[147] M.S. Dresselhaus, G. Dresselhaus and P.C. Eklund, *Science of fullerenes and carbon nanotubes*, Academic Press, San Diego (1996).

[148] P. J. Burke, "Lüttinger Liquid Theory as a Model of the Gigahertz Electrical Properties of Carbon Nanotubes," IEEE. Trans. Nanotechnology, Vol. 1, No. 3, Sept. 2002, pp. 129-144.

[149] M. Bockrath, D. H. Cobden, J. Lu *et al.*, "Luttinger-liquid behaviour in carbon nanotubes," *Nature* Vol. 397, 18 February 1999, pp. 598-601.

[150] C. L. Kane, L. Balents, M. P. A. Fisher, "Coulomb interactions and mesoscopic effects in carbon nanotubes," *Phys. Rev. Lett.* Vol. 79 (1997), pp. 5086–5089.

[151] H.W.C. Postma, *Carbon nanotube junctions and devices*, Doctoral Dissertation, Delft University of Technology, Delft University Press (2001).

[152] P.J. Burke, "A technique to directly excite Luttinger liquid collective modes in carbon nanotubes at GHz frequencies,"
Online Version: http:// xxx.lanl.gov/pdf/cond-mat/0204262

[153] L.D. Landau, "The theory of superfluidity of helium II," *Zh. Eksp. Teor. Fiz.*, Vol. 11, 1941, p. 592.

[154] L.D. Landau, "The theory of superfluidity of helium II," *J. Phys*. USSR, Vol. 11, 91 (1947).

[155] A. J. Leggett, "Superfluidity," *Reviews of Modern Physics*, Vol. 71, No. 2, Centenary 1999, pp. S318-S323.

[156] J. R. Abo-Shaeer, C. Raman, J. M. Vogels *et al.*, "Observation of Vortex Lattices in Bose-Einstein Condensates," *Science*, 20 April 2001, Vol. 292, pp. 476-479.

[157] C. Kittel, *Quantum Theory of Solids*, Second Revised Printing, New York, N.Y.: John Wiley & Sons (1987).

[158] A.A Abrikosov, *Fundamentals of the Theory of Metals*, North-Holland, (1988).

[159] M. Lundstrom, Fundamentals of Carrier Transport, Volume X, Modular Series on Solid State Devices, G.W. Neudeck and R.F. Pierret, Editors, Reading, MA: Addison-Wesley

Publishing (1990).

[160] David R. Tilley, *Superfluidity and Superconductivity*, Institute of Physics Publishing; 3rd edition (October 1, 1990).

[161] H.J. De Los Santos, "On the Design of Photonic Crystal Filters," *1998 USUN/URSI National Radio Science Meeting,* special session on Antennas and Circuits Applications of Photonic Band-Gap Materials, Atlanta, Georgia, June 21-26, 1998.

[162] J.D. Joannopoulos, R.D. Meade, and J. N. Winn, *Photonic Crystals: Molding the Flow of Light*, Princeton, NJ: Princeton University Press (1995).

[163] N. Stefanou, V. Karathanos, ans A. Modinos, "Scattering of electromagnetic waves by periodic structures," *J. Phys.: Condens. Matter*, vol. 4, 1992, pp. 7389-7400.

[164] J.B. Pendry and A. MacKinnon, "Calculations of Photon Dispersion Relations," *Phys. Rev. Lett.*, vol. 69, 1992, pp. 2772-2775.

[165] K. M. Leung and U. Qiu, "Multiple-scattering calculation of the two-dimensional photonic band structure," *Phys. Rev.* B, vol. 48, 1993, pp. 7767-7771.

[166] S.G. Johnson, and J.D. Joannopoulos , *Photonic Crystals: The Road from Theory to Practice*, Boston, MA: Kluwer, Boston (2002).

[167] D. F. Sievenpiper, M. E. Sickmiller, E. Yablonovitch, "3D Wire Mesh Photonic Crystals", *Phys. Rev. Lett* vol. 76, 1996, p. 2480.

[168] Sigalas, M. M., et al., "Metallic Photonic Band-gap Materials", *The American Physical Society*, Vol. 52, No. 16, October 1995, pp. 11744-11751.

[169] V.G. Veselago, Electrodynamics of substances with simultaneously negative electrical and magnetic permeabilities, *Sov. Phys. Usp.*, vol. **10**,1968, p. 509.

[170] J. B. Pendry, "Negative Refraction Makes a Perfect Lens," Phys. Rev. Lett., Vol. 85, No.18, 30 October 2000, pp. 3966-3969.

[171] E.M. Purcell, "Spontaneous Emission Probabilities at Radio Frequencies," *Phys. Rev.*, vol. 69, 1946, p. 681.

[172] D. Kleppner, "Inhibited Spontaneous Emission," *Phys. Rev. Lett.*, Vol. 47, No. 4, 27 July 1981, pp. 233-236.

[173] P. Lodahl, F. van Driel, I.S. Nikolaev et al.,"Controlling the dynamics of spontaneous emission from quantum dots by photonic crystals," *Nature*, Vol. 430, 2004, pp. 654-657.

[174] M. L. Roukes, "Nanoelectromechanical systems," in *Tech. Dig. 2000 Solid-State Sensor and Actuator Workshop*, 2000, p.1..

[175] H.J. De Los Santos, Tracking Analog-to-Digital Converter: US Patent # 05945934, August 31, 1999.

[176] H. Krömmer, A. Erbe, A. Tilkeet al., "Nanomechanical resonators operating as charge detectors in the nonlinear regime," *Europhys. Lett.*, vol. 50, No. 1, pp. 101–106.

[177] A.D. Armour and M.P. Blencowe, "Possibility of an electromechanical which-path interferometer," *Phys. Rev.* B vol. 64, 2001, p. 035311.

[178] H.J. De Los Santos, "Resonator Tuning System," U.S. Patent #6,304.153 B1, issued October 16, 2001.

[179] D.W. Carr, S. Evoy, L. Serakic, *et al.*, "Parametric amplification in a torsional

microresonator," *Appl. Phys. Lett.*, vol. 77, No. 10, 2000, pp. 1545-1547, 2000.

[180] F.M. Serry, D. Walliser, and G.J. Maclay, "The Anharmonic Casimir Oscillator (ACO)—The Casimir Effect in a Model Microelectromechanical System," *J. Microelectromechanical Sys.*, vol. 4, 1995, p.193..

[181] H.B. Chan, V.A. Aksyuk, R.N. Kleiman et al., "Nonlinear Micromechanical Casimir Oscillator," *Phys. Rev. Lett.* vol. 87, 2001, p.211801..

[182] S. Blom, "Magnetomechanics of mesoscopic wires," *Low-Temp. Phys.*, vol. 26, no. 6, 2000, pp. 498-594.

[183] E. Buks and M. Roukes, "Electrically tunable collective modes in a MEMS resonator array," *J. Microelectromech. Syst.*, vol. 11,Dec. 2002, pp. 802-807.

[184] R. Lifshitz and M.C. Cross, "Response of parametrically-driven nonlinear coupled oscillators with application to micro- and nanomechanical resonator arrays," vol. pp. 2003, pp. 134302-134320.

[185] H.J. De Los Santos, "Photonic bandgap crystal frequency multiplexers and pulse blanking filter for use therewith," U.S. Patent 5,749,057, May 5, 1998.

[186] M.O. Scully and M.S. Zubairy, *Quantum Optics*, Cambridge, U.K.: Cambridge Univ. Press (1997).

[187] M. P. Blencowe and M. N. Wybourne, "Quantum squeezing of mechanical motion for micron-sized cantilevers," *Physica B*, vol. 280, 2000, pp. 555-557.

[188] I. Bargatin and M. L. Roukes, "Nanomechanical Laser: Amplification of Mechanical Oscillations by Stimulated Zeeman Transitions," *Phys. Rev. Lett.*, Vol. 91, No. 13, 26 September 2003, pp. 138302-1 -138302-4

[189] C. Bena, S. Vishveshwara, L. Balents *et al.*, "Quantum Entanglement in Carbon Nanotubes," *Phys. Rev. Lett.*, vol. 89, p.037901 (2002)

[190] C. Monroe and D.J. Wineland, "Computing with atoms and molecules," *Science Spectra*, vol. 23, 2000, pp. 72-79.

[191] Cirac, J. I., Zoller, P., "Quantum computation with cold, trapped ions," *Phys. Rev. Lett.* 74, 1995, 4091–4094.

[192] D. J. Wineland, M. Barrett, J. Britton *et al.,* "Quantum information processing with trapped ions," *Phil. Trans. Royal Soc. London* A vol. 361, 2003, pp. 1349-1361.

[193] R. Laflamme, E. Knill, D. G. Cory, *et al.,* " Introduction to NMR Quantum Information Processing," Online Version: http://xxx.lanl.gov/pdf/ arXiv:quant-ph/0207172

[194] L. M.K. Vandersypen and I. L. Chuang, "NMR Techniques for Quantum Control and Computation," Online Version: http://xxx.lanl.gov/pdf/ quant-ph/0404064

[195] M. Steffen, L. M.K. Vandersypen, and I. L. Chuang, " Toward Quantum Computation: A Five-Qubit Quantum Processor," *IEEE Micro*, April 2001, pp. 24-34.

[196] E. M. Purcell, H. C. Torrey, and R. V. Pound. Resonance absorption by nuclear magnetic moment in a solid. *Phys. Rev.*, vol. 69, 1946, pp. 37–38.

[197] F. Bloch, "Nuclear induction," *Phys. Rev.*, vol. 70, 1946 pp. 460–485.

[198] R. R. Ernst, G. Bodenhausen, and A. Wokaun. *Principles of Nuclear Magnetic Resonance in One and Two Dimensions.* Oxford University Press, Oxford (1994).

[199] P. Mansfield and P. Morris, "NMR imaging in medicine," *Adv. Mag. Res.*, 1982, S2:1–343.

[200] D. G. Cory, A. F. Fahmy, and T. F. Havel. Ensemble quantum computing by NMR-spectroscopy, *Proc. Nat. Ac. of Sci. USA*, vol. 94, 1997, pp.1634–1639.

[201] N. A. Gershenfeld and I. L. Chuang. Bulk spin resonance quantum computation. *Science*, vol. 275, 1997, pp. 350–356.

[202] B. E. Kane, "A silicon-based nuclear spin quantum computer," *Nature*, Vol. 393, 14 May 1998, pp. 133-137.

[203] Herring, C. & Flicker, M. Asymptotic exchange coupling of two hydrogen atoms. *Phys. Rev.* vol. 134**,** 1964, pp. A362–A366..

[204] A.O. Caldiera and A.J. Leggett, "Quantum tunneling in a dissipative system," *Ann. Phys. (NY)* 1, vol. 49, 1983, pp. 347-456.

[205] A.J. Leggett, "Testing the limits of quantum mechanics: motivation, state of play, prospects," *J. Phys. CM* vol. 14, 2002, pp. R415-451.

[206] M. H. Devoret†, A. Wallraff, and J. M. Martinis, "Superconducting Qubits: A Short Review, Online Version http://xxx.lanl.gov, cond-mat/0411174

[207] J. M. Martinis, "Superconducting Qubits and the Physics od Josephson Junctions," Online Version: http://xxx.lanl.gov, cond-mat/0402415

[208] H. Mooij, "Superconducting quantum bits," Online Version: http://physicsweb.org/articles/world/17/12/7

[209] Y. Nakamura, Y. A. Pashkin, and J. S. Tsai, "Coherent control of macroscopic quantum states in a single-Cooper-pait box," *Nature*, vol. 398, 29 April 1999, pp. 786-788.

[210] B. E. A. Saleh, and M. C. Teich, *Fundamentals of Photonics*. 1991, New York: Wiley.

[211] S. A. Maier, P. G. Kik, L. A. Sweatlock, et al., "Energy transport in metal nanoparticle plasmon waveguides," *Mat. Res. Soc. Symp. Proc.*, vol. 777, 2003, pp. T7.1.1-T7.1.12.

[212] R. M. Dickson and L. A. Lyon, "Unidirectional Plasmon Propagation in Metallic Nanowires, *J. Phys. Chem. B* 2000, *104, pp.* 6095-6098

[213] J. Takahara, S. Yamagishi, H. Taki et al.,"Guiding of a one-dimensional optical beam with nanometer diameter," *Optics Lett.*, Vol. 22, No. 7, April 1, 1997, pp. 475-477.

[214] B. Lamprecht, J. R. Krenn, G. Schider et al.,"Surface plasmon propagation in microscale metal stripes," *Appl. Phys. Lett.*, Vol. 79, No. 1, 2 July 2001, pp.51-53.

[215] H. Raether, *Surface Plasmons on Smooth and Rough Surfaces and on Gratings*; Springer-Verlag: New York, vol. 111, (1988).

[216] C. J. Barrelet, A. B. Greytak, and C. M. Lieber, "Nanowire Photonic Circuit Elements," *Nano Lett.*, Vol. 4, No. 10, 2004, pp. 1981-1985.

[217] T. Klar, M. Perner, S. Grosse et al., "Surface-Plasmon Resonances in Single Metallic Nanoparticles," Vol. 80, No. 19, *Phys. Rev. Lett.*, 11 May, 1998.

[218] S. Link and M. A. El-Sayed, "Shape and size dependence of radiative, non-radiative and

photothermal properties of gold nanocrystals," *Int . Reviews inn Physical Chemistry* , 2000, Vo l . 19, No . 3, pp.409

[219] J. R. Krenn, A. Dereux, J. C. Weeber et al., "Squeezing the Optical Near-Field Zone by Plasmon Coupling of Metallic Nanoparticles," *Phys. Rev, Lett.*, Vol. 82, No. 12, 22 March 1999, pp. 2590-2593.

[220] J. P. Kottmann and O. J.F. Martin. "Plasmon resonant coupling in metallic nanowires,"

Optics Express, Vol. 8,

[221] S. I. Bozhevolnyi, J. Erland, K. Leosson et al., "Waveguiding in Surface Plasmon Polariton Band Gap Structures," *Phys. Rev. Lett.*, vol. 26, No. 14, 2 April 2001, pp. 3008-3011.

[222] J. R. Krenn, H. Ditlbacher, G. Schider et al., "Surface plasmon micro- and nano-optics," *Journal of Microscopy, Vol. 209, Pt 3 March 2003, pp. 167–172*

[223] M. H. Huang, S.l. Mao, H. Feick *et al.,* "Room-Temperature Ultraviolet Nanowire Nanolasers," *Science*, Vol. 292, 8 June 2001, pp. 1897-1899.

[224] D. W. Pohl, "Near-Field Optics and the Surface Plasmon Polariton," Chapter in S. Kawata (Ed.): *Near-Field Optics and Surface Plasmon Polaritons, Topics Appl. Phys.*, vol. 81, Berlin: Springer-Verlag (2001), pp. 1-13.

[225] R. C. Dunn, "Near-Field Scanning Optical Microscopy," *Chem. Rev., Vol. 99,* 1999, pp. 2891-2927

[226] E.H. Synge, "A suggested method for extending the microscopic resolution into the ultramicroscopic region" *Phil. Mag.* Vol. 6, 356 (1928).

[227] E.H. Synge, "An application of piezoelectricity to microscopy", *Phil. Mag.*, vol. 13, 1932, p. 297

[228]Online Version: http://www.nanonics.co.il/main/twolevels_item1.php?ln=en&item_id=34&main_id=14

[229] H. A. Bethe, H. A. "Theory of diffraction by small holes," *Phys. Rev.*, vol. *66*, 1944, pp. 163-182.

[230] CJ Bouwkamp, "On the diffraction of electromagnetic waves by small circular disks and holes," Philips Res. Rep. vol. 5, 1950, pp. 401-422.

[231] CJ Bouwkamp, "On Bethe's theory of diffraction by small holes," Philips Res. Rep., vol. 5, 1950, pp. 321–332.

[232] D. S. Koltun and J. M. Eisenberg, *Quantum Mechanics of Many Degrees of Freedom*, New York, N.Y.: John Wiley & Sons, Inc.,(1988).

[233] S. Raimes, *Many-Electron Theory*, New York, N.Y.: American Elsevier Publishing Co., (1972).

INDEX